AGRICULTURE ISSUES AND POLICIES

SORGHUM

PRODUCTION, GROWTH HABITS AND HEALTH BENEFITS

AGRICULTURE ISSUES AND POLICIES

Additional books in this series can be found on Nova's website under the Series tab.

Additional e-books in this series can be found on Nova's website under the e-book tab.

FOOD SCIENCE AND TECHNOLOGY

Additional books in this series can be found on Nova's website under the Series tab.

Additional e-books in this series can be found on Nova's website under the e-book tab.

AGRICULTURE ISSUES AND POLICIES

SORGHUM

PRODUCTION, GROWTH HABITS AND HEALTH BENEFITS

PATRICIA C. PARRA
EDITOR

NOVINKA

New York

For permission to use material from this book please contact us:
Telephone 631-231-7269; Fax 631-231-8175
Web Site: http://www.novapublishers.com

NOTICE TO THE READER

Library of Congress Cataloging-in-Publication Data

Sorghum : production, growth habits and health benefits / editor: Patricia C. Parra.
p. cm.
Includes index.
ISBN 978-1-62618-351-3 (soft cover)
1. Sorghum. 2. Sorghum--Utilization. I. Parra, Patricia C.
SB191.S7S666 2013 633.6'2--dc23 2013007545

Published by Nova Science Publishers, Inc. † New York

CONTENTS

PREFACE

Sorghum is the fifth most important grain crop after wheat, rice, maize and barley. It is cultivated for food and beverages for humans, and for feed and fodder for animals. In this book, the authors study the production, growth habits and health benefits of sorghum. Topics discussed include sorghum as an enigmatic grain for chicken-meat production; the potential of sorghum for development in East Africa; the health benefits of sorghum grain for poultry and pigs; fungal root endophytes as sorghum growth promoters; and breeding for abiotic stresses in sorghum.

Chapter 1 - In Australia, if not other countries, sorghum is an enigmatic grain for chicken-meat production because the performance of broiler chickens on sorghum-based diets is considered to be either occasionally or routinely inferior to those offered wheat-based diets. This chapter considers a number of factors which may be contributing to this situation including pellet quality, protein quality, grain texture, particle size of sorghum following hammer-milling and conditioning temperatures at which sorghum-based diets are steam-pelleted. However, it is considered that the challenges posed by these factors can be addressed, at least partially, and the appropriate strategies are discussed. The authors' contention is that the fundamental problem with sorghum is that its energy density, most of which is derived from starch, is relatively poorly utilised by broiler chickens and the underlying reasons have not been clarified. Considerable attention has been paid to the possible limiting role of kafirin, the dominant protein fraction in sorghum, on starch gelatinisation and digestibility. However, other factors that may be impeding starch digestion include glutelin, the second protein fraction in sorghum, phytate, and phenolic compounds (including 'non-tannin' phenolics, ferulic acid and possibly condensed tannin). It is proposed that the identification of the relative importance of these factors on sorghum starch digestion is key to solving the conundrum that is sorghum grain in chicken-meat production.

Chapter 2 - Sorghum domestication probably dates back to the advent of agriculture in sub-Saharan Africa ~10,000 years ago. Today, it is an established understanding that the Sub-Saharan and North East regions of Africa are the primary centers of sorghum diversity. On the African continent, sorghum is grown on 25 million hectares, which constitutes approximately 61% of the world total however African yields average only 0.85 t/ha compared with 4.5 t/ha in North America. The low productivity is associated with; traditional modes of production, low levels of technology adoption, low uptake of improved varieties and hybrid-seed technology, environmental factors such as unpredicted drought periods and other soil-edaphic constraints, institutional bottlenecks such as research capacity, information barriers, poor infrastructure, marketing hardships, and environmental stresses the major determinants of present sorghum production and yield losses. The authors have cloned resistance genes (NB-LRR type) to the disease inciting fungi *Setosphaeria turcica* (*St* genes) and *Colletotrichum sublineolum* (*Cs* genes) causing turcicum leaf blight and sorghum anthracnose, respectively, using a differential transcriptomic approach on Ugandan sorghum genotypes. Presently the authors are re-sequencing different sorghum genotypes to enhance the authors' understanding of sorghum speciation but more importantly to generate knowledge on additional traits of interest. Wild sorghum species can survive with less water uptake, resist more diseases and insect pests and are presently used to widening the genetic diversity in a pre-breeding initiative. Further, molecular markers are being developed for a genomic selection approach in order to facilitate and speed up the selection processes not least to meet the climate change and emergent food security constrains due to the rapid population growth.

Chapter 3 - Sorghum is the fifth most important grain crop after wheat, rice, maize and barley. Sorghum is cultivated for food and beverages for humans, and for feed and fodder for animals in America, Asia, Australia and Africa. Its crude protein content is higher than that of maize whereas its energy value is rated as high as that of maize. Sorghum tolerates poorer growing conditions than maize, so is more abundant in marginal regions, where it is used to feed humans and animals. Sorghum may confer some health benefits on consumers due to the nature of its carbohydrates but also the presence of other chemicals, which ordinarily are regarded as antinutrients in animal nutrition, e.g. polyphenols or tannins and polycosanols. When fed to humans, these factors may be beneficial. Tannins would reduce feed intake and digestive enzyme activities, to minimize weight gain while polycosanols, present in the waxy layer, are known to reduce the concentrations of low-density lipoproteins and improve cardiac function. Sorghum is also free from gluten and therefore useful for people suffering from coeliac disease. Some of these benefits are also noticeable in animals although could be negative to productivity. Sorghum is therefore not seen to be an ingredient for controlling animal health and its use as such has not received significant attention in animal nutrition. This chapter discusses the direct health benefits associated with feeding sorghum grains to poultry and pigs. It will also link changes in animal product quality to health benefits of the human consumer.

Chapter 4 – Fungal endophytes are found in healthy plant tissues during at least one stage of their life cycle and do not cause any apparent symptoms of disease or negatively affect the host. These fungi have been isolated from all plants studied in natural ecosystems. They can induce many changes in the plant ecology, fitness and evolution and, thus, can influence whole plant communities. The Ascomycetous root fungal endophytes can be divided into two groups, the dark septate endophytes (DSE) and the fungi with hyaline and pale hyphae. Fungal endophyte colonization is important for the improvement of the ecological adaptability of the host, resulting in enhanced tolerance of both biotic and abiotic stresses. Another important group of endophytes are those that constitute arbuscular mycorrhizal (AM) symbiosis. AM symbiosis is very important for plant nutrient uptake and growth and for plant resistance to biotic and abiotic stresses. It is known that AM symbiosis can be influenced by other root fungal endophytes, which confer benefits to the host plant, such as growth promotion, protection against disease or enhanced phosphorous uptake. Despite the benefits of these root fungal endophytes for agronomical plants, the physiological and ecological significance of these fungi remains poorly characterized. In this context, the authors discuss the potential roles of fungal endophytes for growth promotion in *Sorghum*.

Chapter 5 - Mapping and analysis of QTL controlling tolerance to abiotic stresses are primary and significant approaches for germplasm enhancement in crops. In particular, corresponding to global climate changes, improved tolerance to abiotic stresses contributes to crop adaptation and production under stress conditions in crops. The tolerance to abiotic stresses is frequently controlled by multiple genes or quantitative trait loci (QTLs). Identification of QTL using DNA markers is the first strategy in DNA marker-assisted selection (MAS) for the tolerance in sorghum breeding. The authors assessed salt and drought tolerances using a core collection derived from world germplasm collections of sorghum and detected a broad range of variation.

Association analysis was performed using data on genotypes at 98 SSR marker loci and phenotypic values. Major QTL for salt tolerance was mapped on chromosome 1 and closely linked to QTL for leaf senescence. Nine QTL for drought tolerance were also detected. Based on the genetics of tolerance to abiotic stresses, the authors discussed breeding strategies for abiotic stresses in sorghum.

In: Sorghum
Editor: Patricia C. Parra

ISBN: 978-1-62618-351-3

Chapter 1

SORGHUM: AN ENIGMATIC GRAIN FOR CHICKEN-MEAT PRODUCTION

Peter H. Selle*[*], *Sonia Y. Liu and Aaron J. Cowieson
Poultry Research Foundation within the University of Sydney, Australia

ABSTRACT

In Australia, if not other countries, sorghum is an enigmatic grain for chicken-meat production because the performance of broiler chickens on sorghum-based diets is considered to be either occasionally or routinely inferior to those offered wheat-based diets. This chapter considers a number of factors which may be contributing to this situation including pellet quality, protein quality, grain texture, particle size of sorghum following hammer-milling and conditioning temperatures at which sorghum-based diets are steam-pelleted. However, it is considered that the challenges posed by these factors can be addressed, at least partially, and the appropriate strategies are discussed. Our contention is that the fundamental problem with sorghum is that its energy density, most of which is derived from starch, is relatively poorly utilised by broiler chickens and the underlying reasons have not been clarified. Considerable attention has been paid to the possible limiting role of kafirin, the dominant protein fraction in sorghum, on starch gelatinisation and digestibility. However, other factors that may be impeding starch digestion include glutelin, the second protein fraction in sorghum, phytate, and phenolic compounds (including 'non-tannin' phenolics,

[*] Corresponding author: peter.selle@sydney.edu.au.

ferulic acid and possibly condensed tannin). It is proposed that the identification of the relative importance of these factors on sorghum starch digestion is key to solving the conundrum that is sorghum grain in chicken-meat production.

1. Introduction

The annual consumption of chicken-meat in Australia is high, in the order of 44 kg per capita, and this demand is met by the processing of 512 million broiler chickens for an annual output of 934,000 tonnes of chicken-meat (ACMF, 2011). While wheat is the dominant grain, sorghum makes up approximately one-third of the cereal base of Australian broiler diets as either sorghum *per se* or as wheat-sorghum blends. The Australian sorghum harvest has averaged 2.236 million tonnes from 2008 to 2012 inclusive and the entire crop could be utilised as an animal feedstuff for pigs, poultry and feedlot cattle. However, up to 500,000 tonnes of sorghum may be exported annually, which reflects a certain reluctance by these primary industries to incorporate sorghum into their diets despite the fact that sorghum is usually available at a discount price relative to wheat.

In Australia, the performance of broiler chickens offered sorghum-based diets has been described as sub-optimal (Bryden et al., 2009b) and considerable efforts have been made to identify the underlying, causal factors. This is evidenced by two lengthy reports (Bryden et al. 2009b; Perez-Maldonado and Rodrigues, 2009) and several reviews (Gidley et al., 2011a,b; Selle et al., 2010a; Selle, 2011). However, our view is that "inconsistent" is probably a better description of broiler performance on sorghum-based diets than "sub-optimal". In a sorghum project recently completed by Poultry Research Foundation weight gains of broiler chickens ranged from clearly inferior in one study (Selle et al., 2013a), relative to 2012 Ross 308 performance objectives, to somewhat superior in another study (Selle et al., 2012). Also, in an earlier study (Selle et al., 2010b), sorghum-based diets supported numerically better 42-day growth performance (weight gain: 2507 versus 2462 g/bird; FCR: 1.845 versus 1.929) than wheat-based diets without the addition of feed enzymes. However, the dietary inclusions of either xylanase or an enzyme cocktail (protease, β-glucanase, xylanase) resulted in nearly identical feed conversion ratios (1.839 versus 1.838) so that the 4.35% advantage in FCR for sorghum was eroded by exogenous enzymes. Predictably, this indicates that 'viscous' wheat-based diets were more responsive to the inclusion of NSP-degrading enzymes than 'non-viscous' sorghum-based diets. The routine inclusion of xylanase in wheat-based broiler diets is an advantage not shared by sorghum and probably contributes to the perception that sorghum is sub-optimal.

An informal survey of 15 practical, 'coal-face' nutritionists was completed to seek their opinions as to the performance of pigs and poultry on sorghum-based diets and to indentify the important factors inherent in sorghum. The nutritionists were asked to compare the performance of pigs and poultry on sorghum-based diets relative to wheat plus xylanase. A clear majority (71%) believed that performances on sorghum-based diets were either routinely or occasionally inferior while a minority (29%) considered sorghum and wheat to be equivalent. The predominant view (57%) was that sorghum supported inconsistent (occasionally inferior) performances in pigs and poultry. However, the dichotomy between the 29% of respondents who believe sorghum performance is equivalent to wheat and the 14% who consider that consider sorghum is *consistently* inferior (or sub-optimal) to wheat illustrates the enigmatic nature of grain sorghum and this chapter seeks to address the conundrum that is sorghum in chicken-meat production.

2. Pellet Quality of Sorghum-Based Diets

An interesting outcome was that while no reference to pellet quality was made in the survey questionnaire, a number of respondents volunteered the information that they held concerns about the pellet quality of sorghum-based diets. This is not surprising as the 61°C starch gelatinisation temperature of wheat is considerably lower than 73°C of sorghum (Taylor and Dewar, 2001). Consequently, in order to achieve adequate starch gelatinisation and acceptable pellet quality sorghum-based diets are usually steam-pelleted at high conditioning temperatures, often in excess of 90°C, but the concerns about pellet quality remain.

Poor pellet quality, and the presence of fines, reduces voluntary feed intakes and therefore weight gain. For example, in one study (Selle et al., 2013a) involving two sorghum varieties and conditioning temperatures of 65 and 97°C, intact pellets were reground to mash with, effectively, "zero" pellet quality. This transition reduced feed intake by 8.95% (1903 versus 2090 g/bird; $P < 0.001$) and weight gain by 8.32% (1223 versus 1334 g/bird; $P < 0.001$) from 7 to 28 days post-hatch, which illustrates the importance of pellet integrity to voluntary feed intakes.

It is generally accepted that the extent of starch gelatinisation generated by the steam-pelleting process, which is usually in the order of 20% and relatively modest, is still an important driver of pellet quality. Therefore, it is relevant that in the Chandrashekar and Kirleis (1988) study, the average proportion of kafirin, the dominant protein fraction in sorghum, was 54.9% of total protein across nine sorghum varieties. However, the kafirin proportion of protein was inversely related to the degree of starch gelatinisation ($r = -0.683$; $P < 0.05$), which suggests that kafirin was impeding starch gelatinisation. Moreover, a reducing agent was evaluated and 2-mercaptoethanol significantly ($P < 0.005$) increased the extent of starch gelatinisation in the nine sorghums from 79.1 to 95.2%. Reducing agents have the capacity to cleave disulphide bonds and those located in the periphery of kafirin protein bodies (β- and γ-kafirin) are believed to hold particular importance (Duodu et al., 2004).

The suggestion arising out of the Chandrashekar and Kirleis (1988) data is that the kafirin proportion of sorghum protein hinders starch gelatinisation and presumably compromises pellet quality as a result. It follows that more acceptable pellet quality may be achieved with diets based on 'low-kafirin' sorghums. There is, however, the real possibility that proportions of kafirin in sorghum have increased over the last decade in Australia, perhaps as an unintended consequence of breeding programs (Selle, 2011).

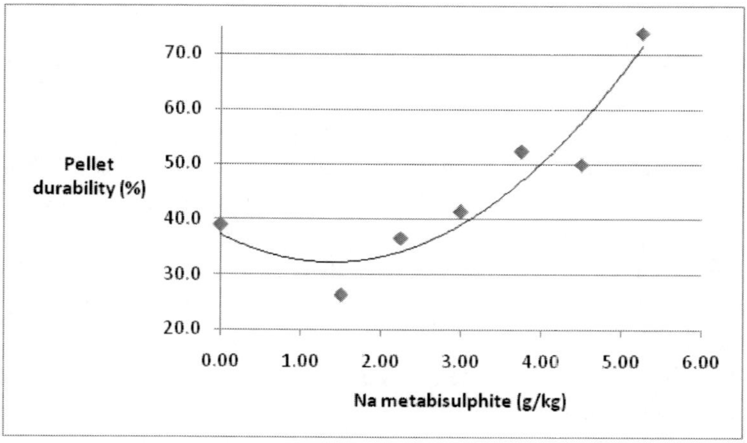

Figure 1. The effect of sodium metabisulphite dietary inclusion rates in sorghum-based diets on pellet durability (mean of 1, 2 and 3 minutes) [Quadratic effect: $r = 0.947$; $P = 0.011$].

Table 1. Concentrations and apparent ileal digestibility (AID) coefficients of amino acids in 17 sorghum samples with crude protein contents ranging from 71 to 118 g/kg (mean: 101.9 ± 14.35 g/kg) and linear regressions between crude protein and amino acid concentrations (adapted from Bryden et al., 2009a)

Amino acid	Concentration (g/kg)			AID coefficient		Regression	
	Mean	Range	C of V (%)	Mean	Range	r =	P =
Arginine	3.72	2.5-5.1	21.4	0.795	0.71-0.86	0.390	0.122
Histidine	2.29	1.7-3.1	17.2	0.717	0.62-0.80	0.649	0.005
Isoleucine	4.12	2.7-5.4	23.2	0.818	0.74-0.87	0.639	0.006
Leucine	13.57	9.2-17.0	19.9	0.855	0.79-0.92	0.721	0.001
Lysine	2.07	1.4-2.6	20.3	0.743	0.67-0.83	0.329	0.197
Methionine	1.57	1.1-2.1	24.0	0.818	0.75-0.88	0.399	0.113
Phenylalanine	5.23	3.4-6.7	21.9	0.835	0.75-0.90	0.630	0.007
Threonine	3.16	2.4-4.3	17.0	0.678	0.58-0.76	0.725	0.001
Valine	5.04	3.4-6.5	21.3	0.788	0.71-0.85	0.614	0.009
Alanine	9.11	6.1-11.9	21.8	0.856	0.79-0.92	0.698	0.002
Aspartic acid	6.49	4.3-8.5	21.4	0.798	0.72-0.87	0.671	0.003
Glutamic acid	20.85	14.6-26.3	19.6	0.846	0.73-0.90	0.735	0.001
Glycine	3.06	2.1-4.1	19.9	0.721	0.64-0.92	0.599	0.020
Serine	4.51	3.5-6.7	20.2	0.771	0.70-0.84	0.665	0.004
Tyrosine	3.25	1.9-5.4	30.5	0.775	0.71-0.84	0.322	0.208

Recently, the Poultry Research Foundation has been evaluating the inclusion of the reducing agent sodium metabisulphite in sorghum-based broiler diets. Initially, it was not intended to investigate the possible effects of sodium metabisulphite on starch gelatinisation and pellet quality. However, in light of the above, the impact of graded sodium metabisulphite inclusion levels in conventional sorghum-based diets on their pellet durability index has been assessed and the extent of starch gelatinisation will be determined. The results of the initial pellet integrity investigation are encouraging. Sodium metabisulphite quadratically ($P < 0.012$) increased pellet durability indices with positive responses being observed at inclusion rates ranging from 3.00 to 5.25 g/kg (Figure 1). At 5.25 g/kg, sodium metabisulphite increased the pellet durability index from 39.0% to 74.0%.

It may be overly simplistic to associate only starch gelatinisation with enhanced pellet quality, as Wood (1987) reported that raw protein had a greater impact on pellet quality (durability and hardness) than pre-gelatinised starch. Also, Briggs et al. (1999) contend that protein has an important role in pellet quality and suggest the role of starch gelatinisation in respect of pellet durability is over-stated. However, the impact of sodium metabisulphite on starch gelatinisation should establish whether the improvements in pellet quality observed were a response to enhanced starch gelatinisation or not. If not, it is possible that the direct impact of sodium bisulphite on dietary proteins was a more important factor. Nevertheless, it would be an important outcome if it is confirmed that reducing agents can enhance pellet quality.

3. Protein Quality of Sorghum

Considerable emphasis has been placed on the relatively poor protein quality of sorghum as a causal factor for its real or perceived shortcomings as a feedstuff in broiler diets with a particular focus on the kafirin component of sorghum protein. However, in a typical sorghum-based broiler diet the amount of total protein derived from sorghum is in the order of 30% and the dietary inclusion of sorghum is primarily as a source of energy, the majority of which is derived from starch.

From data generated by Bryden et al. (2009a), concentrations and apparent ileal digestibility (AID) coefficients of 15 amino acids in 17 Australian sorghum varieties are shown in Table 1. The overall mean AID coefficient was 0.788, which ranged from 0.678 (threonine) to 0.855 (methionine), indicating that sorghum amino acids are not highly digestible. This data is in agreement with that of Perez-Maldonado and Rodrigues (2009) who reported that in 14 sorghums from the 2005 harvest the mean AID coefficient of 12 amino acids was 0.768 while threonine and methionine had AID coefficients of 0.681 and 0.853, respectively (Table 2). However, in the previous 2004 harvest the mean AID coefficient of 17 sorghums was 0.651, with 0.527 for threonine and 0.759 for methionine. This represents a notable reduction of 15.2% in average AID coefficients and emphasises the variability in sorghum amino acid digestibilities. Moreover, as shown in Table 1, the concentrations of arginine ($P > 0.12$), lysine ($P > 0.19$), methionine ($P > 0.11$) and tyrosine ($P > 0.20$) are not significantly correlated to the crude protein contents of the 17 sorghums, which ranged from 71 to 118 g/kg around a mean value of 101.9 g/kg. Clearly, concentrations of the balance of amino acids were significantly correlated to crude protein levels but both lysine and methionine are critical amino acids. Thus both the concentrations and ileal digestibilities of amino acids in sorghum are variable, which constitutes a fundamental challenge.

There are four fractions of sorghum protein; Virupaksha and Sastry (1968) reported that kafirin constituted 54.1%, glutelin 33.4%, globulin 7.0% and albumin 5.6% of endosperm protein in five sorghum samples. Of the dominant fractions, kafirin is located in protein bodies and glutelin comprises the protein matrix of sorghum endosperm (Seckinger and Wolf, 1973), where both kafirin and glutelin are intimately associated with starch granules.

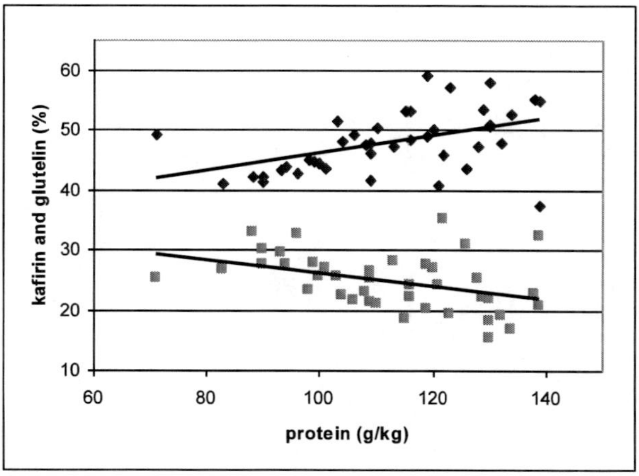

Figure 2. Relationships between sorghum protein concentrations and kafirin ($r = 0.886$; $P < 0.005$) and glutelin ($r = -0.401$; $P < 0.01$) as proportions (%) of protein (adapted from Taylor et al. 1984).

Table 2. Concentrations (g/kg) and apparent ileal digestibility (AID) coefficients of amino acids in sorghum samples harvested in 2004 and 2005 with percentage differences in AID between harvests (adapted from Perez-Maldonado and Rodrigues, 2009)

Amino acid	2004 harvest (n = 17)			2005 harvest (n = 14)	% difference 2005 vs 2004
	g/kg	AID	g/kg	AID	
Crude protein	116.5		115.9		
Arginine	4.36	0.695	4.15	0.805	15.8
Histidine	1.64	0.581	2.18	0.727	25.1
Isoleucine	4.16	0.683	3.93	0.805	17.9
Leucine	14.26	0.739	13.64	0.849	14.9
Lysine	2.97	0.707	1.87	0.769	8.8
Methionine	1.39	0.759	1.62	0.853	12.4
Phenylalanine	4.14	0.728	5.21	0.833	14.4
Threonine	2.91	0.527	3.19	0.681	29.2
Tryptophan	1.41	0.584	1.29	0.777	33.0
Valine	5.18	0.662	4.80	0.784	18.4
Cystine	1.58	0.452	1.93	0.531	17.5
Tyrosine	2.98	0.690	3.82	0.806	16.8
Sum/Mean	46.98	0.651	47.63	0.768	18.7

Table 3. The amino acid profiles (g amino acid /16 g N) of sorghum protein and kafirin and glutelin fractions of sorghum protein [adapted from Jambunathan and Mertz, (1973)[1]; Bryden et al., (2009a)[2]; Perez-Maldonado and Rodrigues, (2009)[3]; Salunkhe et al., (1977)[4] ; Mosse et al., (1988)[4]]

Amino acid	Sorghum			Kafirin[4]	Glutelin[4]
	A[1]	B[2]	C[3]		
Arginine	3.48	3.69	3.68	1.06	4.91
Histidine	1.95	2.26	1.62	0.70	3.60
Isoleucine	3.96	4.04	3.48	4.00	4.40
Leucine	13.85	13.32	11.99	17.37	10.40
Lysine	2.08	2.06	2.15	0.23	2.94
Methionine	1.00	1.55	1.29	0.89	1.30
Phenylalanine	5.04	5.14	3.96	5.63	4.57
Threonine	3.16	3.12	2.50	1.87	2.40
Tryptophan	-	-	1.17	0.18	1.07
Valine	5.19	4.95	4.31	5.14	7.04
Alanine	9.60	8.93	-	9.60	9.01
Aspartic acid	7.60	6.37	-	5.08	5.57
Cystine	0.89	-	1.49	0.54	1.16
Glutamic acid	22.54	20.47	-	27.33	21.47
Glycine	2.98	3.02	-	1.67	3.43
Proline	7.92	-	-	10.71	8.96
Serine	4.39	4.44	-	3.56	5.06
Tyrosine	4.37	3.21	2.87	4.50	3.06

Subsequently, Taylor et al. (1984) reported average concentrations of 54.0 g/kg kafirin and 27.1 g/kg glutelin in 41 sorghum samples with an average crude protein content of 111 g/kg. Thus kafirin and glutelin comprised 48.0% and 27.7% of sorghum protein, respectively. However, kafirin, as a proportion of protein, was positively correlated ($r = +0.469$; $P < 0.005$) with sorghum protein concentrations; whereas, the proportion of glutelin was negatively correlated ($r = -0.401$; $P < 0.01$) as shown Figure 2. Thus, the proportion of kafirin increased at the expense of glutelin with increasing sorghum protein levels. Importantly, however, the amino acid profiles of kafirin and glutelin are distinctly different as shown in Table 3 from composite data generated by Salunkhe et al. (1977) and Mosse et al. (1988). Glutelin contains considerably more basic amino acids (arginine, histidine, lysine) than kafirin and also contains higher levels of tryptophan and cystine; whereas, kafirin contains considerably more leucine than glutelin. Consequently, the reciprocal nature of kafirin and glutelin concentrations in sorghum influences the overall amino acid composition of the grain and this almost certainly contributed to the non-significant relationships between crude protein levels and concentrations of arginine, lysine, methionine and tyrosine in the sorghum data generated by Bryden et al. (2009a).

The Landry and Moreaux (1970) method is the classical approach to quantify kafirin concentrations in sorghum. From a review (Selle et al., 2010a) of six studies, the average kafirin proportion of protein ranged from 37.5 to 72.9% around a mean of 49.3% across 88 sorghum samples. Hamaker et al. (1995) reported a range of kafirin proportions from 49.0 to 54.7% in five sorghums using the classical method but a higher range of 68.1 to 72.9% using the alternative method of Wallace et al. (1990). Quantifying the variable concentration of kafirin in sorghum is not straightforward and this is a real barrier to defining its nutritive importance.

There are three classifications of kafirin where α-kafirin occupies the central core of the protein body while β- and γ-kafirin are located in the periphery. As reviewed by De Mesa-Stonestreet et al. (2010), α-kafirin is dominant (66-84%) relative to β- kafirin (8-13%) and γ-kafirin (9-21%); however, quite different proportions have been reported (Selle et al., 2010a). The amino acid composition of the three kafirin categories as determined by Watterson et al. (1990) and Shull et al. (1992) are shown in Table 4, where lysine levels are uniformly low, leucine is highest in α-kafirin and cystine levels are noticeably higher in β- and γ-kafirin.

Table 4. Amino acid profiles (% total amino acids) of sorghum kafirin fractions in sorghum (adapted from Watterson et al., 1990; Shull et al., 1992)

Amino acid	α-kafirin	β-kafirin	γ-kafirin
Arginine	1.12	3.80	2.84
Histidine	1.50	1.13	8.71
Isoleucine	4.64	2.44	2.78
Leucine	15.92	12.73	9.61
Lysine	0.47	0.59	0.48
Methionine	0.72	6.88	1.21
Phenylalanine	5.31	2.54	2.42
Threonine	3.07	4.43	4.07
Valine	4.52	4.93	5.53
Alanine	10.89	9.66	4.57
Aspartic acid	6.00	3.55	0.65
Cystine	1.07	4.80	6.80
Glutamic acid	25.80	21.06	16.53
Glycine	2.48	4.13	5.26
Proline	8.24	9.03	21.18
Serine	4.90	3.91	4.28
Tyrosine	3.35	4.40	3.10

Glutelin has received less attention than kafirin; however, Beckwith (1972) reported that glutelin accounted for more than half the total protein in three sorghum hybrids and glutelin contained more cystine and basic amino acids than kafirin. Further, Beckwith (1972) proposed that glutelin consisted of small protein units linked by disulphide bridges and, importantly, the reducing agent β-mercaptoethanol substantially increased glutelin solubility from 23 to 87%.

Therefore, both kafirn:glutelin ratios, and the constituency of kafirin, are determinants of the amino acid composition of sorghum. The high leucine content in kafirin is noteworthy as excess leucine has been shown to compromise isoleucine and valine availability in poultry and leucine may also have the capacity to depress voluntary feed intakes (Morrison et al., 2007). Alternatively, the low lysine content in kafirin is noteworthy and the relative amount of lysine in sorghum declines with increasing protein contents ($r = -0.880$; $P < 0.001$) as shown in Figure 3 from data generated by Mosse et al. (1988). This is entirely consistent with an increase in proportionate amounts of kafirin with increasing protein levels deduced from the Taylor et al. (1984) study.

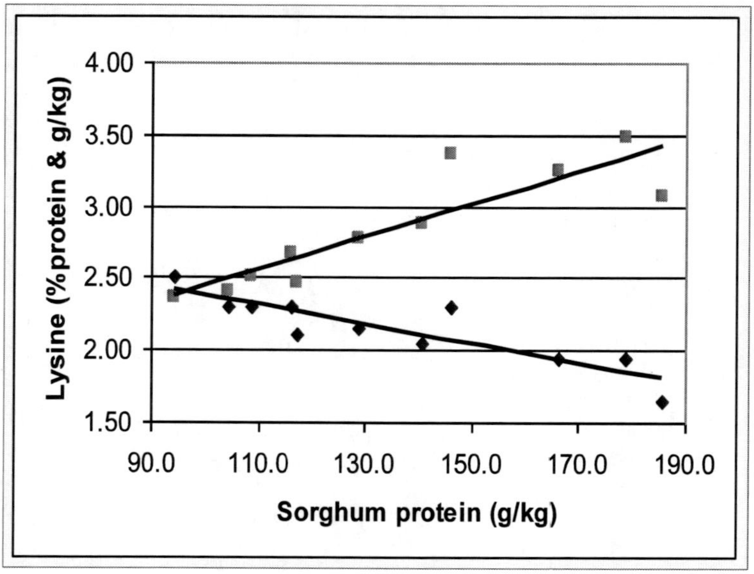

Figure 3. The significant ($P < 0.001$) linear relationships between sorghum protein and lysine in absolute (g/kg) and relative (% protein) terms, where $r = 0.887$ and $r = -0.880$, respectively (adapted from Mosse *et al.* 1988).

Initially, Ravindran et al. (1998) surveyed the concentration and digestibility of amino acids in six local sorghums and subsequently Bryden et al. (2009a) analysed a further eleven samples; the combined results are summarised in Table 1. It is possible to compare data from the two surveys where the mean protein content of the six "1998 sorghums" (98.7 g/kg) and the eleven "2009 sorghums" (103.6 g/kg) were quite similar and there were no significant differences in AID coefficients between the two sub-sets. However, the 2009 sorghums contained significantly less lysine (1.89 versus 2.40 g/kg; $P < 0.015$) and arginine (3.32 versus 4.47 g/kg; $P < 0.005$) than the 1998 sorghums. Given that kafirin contains a paucity of lysine and arginine these reduced concentrations suggest that kafirin, as a proportion of total sorghum protein, was greater in 2009 than in 1998. Thus this comparison raises the possibility that kafirin:glutelin ratios in Australian sorghums have increased over the past decade, which may be an unintended consequence of breeding programs.

Clearly, it is disadvantageous if the kafirin proportion of sorghum protein is indeed increasing in part because kafirin contains low levels of lysine and high levels of leucine. Moreover, kafirin is a poor source of digestible amino acids due to its inherent hydrophobicity and the structure of protein bodies. It is generally accepted that disulphide cross-linking in the β- and γ-kafirin fractions, which are located in the periphery of the protein body, impedes the digestion of the central α-kafirin core and that this is exacerbated by hydrothermal processes (eg 'wet-cooking') because they induce disulphide bond formation.

Quite recently, there has been an increasing focus on exogenous proteases as their inclusion in pig and poultry diets may enhance the digestibility of amino acids. Taylor (2005) suggested that proteases should be evaluated in sorghum-based broiler diets with the caveat that kafirin and glutelin are not readily hydrolysed by proteolytic enzymes. Nevertheless, Sultan et al. (2010, 2011a,b,c) have reported a series of promising results in N digestibility following the inclusion of various feed enzymes including a protease in 'all-sorghum' diets. Moreover, Selle et al. (2013b) reported that a *Bacillus lichenformis*-derived protease (Cibenza^{TM} DP100; Novus international Inc.) improved N digestibility in conventional, sorghum-based broiler diets that were steam-pelleted at an 80°C conditioning temperature. At the standard inclusion rate, this protease significantly increased N digestibility coefficients by 16.5% in the distal jejunum (0.627 versus 0.538; $P < 0.005$), by 5.91% in the proximal ileum (0.770 versus 0.727; $P < 0.05$) and by 7.09% in the distal ileum (0.770 versus 0.719; $P < 0.005$). The Poultry Research Foundation has evaluated two other exogenous proteases in sorghum-based diets but the responses observed did not hold equal promise.

Consequently, Liu et al. (2013a) determined the effect of the *Bacillus lichenformis*-derived protease on the apparent digestibility of amino acids in four small intestinal sites and the results are presented in Table 5. In essence, the main effect of protease was to improve significantly ($P < 0.05$ - < 0.001) the digestibility of arginine (4.63%), histidine (8.23%), isoleucine (9.87%), leucine (12.7%), phenylalanine (10.5%), threonine (7.36%), valine (10.4%), alanine (12.8%), aspartic acid (9.60%), glutamic acid (9.38%), glycine (9.73%), proline (14.6%), serine (11.5%), and tyrosine (12.1%) where the marked percentage improvements are shown in parentheses. Only the responses in lysine and methionine digestibility were not significant and this was probably because the diets contained quite high levels of these two amino acids in 'free' or synthetic forms, which are highly digestible. In addition, protease supplementation significantly ($P < 0.05$ - < 0.005) increased the rates of digestion of 12 of the 16 amino acids assessed. The significant percentage increases in rates of digestion ranged from 34.2% for histidine to 55.7% for both proline and tyrosine. Moreover, as proposed by Liu et al. (2013b), it is probable that digestion rates of amino acids are more closely related to broiler performance than amino acid digestibility coefficients.

4. The Utilisation of Energy Derived from Sorghum (Starch)

While sorghum may provide 30% of protein in a sorghum-based broiler diet in the order of 60% of its energy density is derived from the starch component of sorghum. However, Black et al. (2005) expressed concerns about the adequacy with which the starch/energy component of sorghum is utilised by broiler chickens. These concerns were also reflected in the survey of practical nutritionists. 39% of respondents nominated energy (starch) utilisation outright when asked to identify the reasons why sorghum may be inferior to wheat. However, 46% nominated both starch and protein utilisation because the two factors may be interdependent but not one respondent attributed the shortcomings of sorghum solely to protein utilisation.

Much of the current thinking in relation to the digestibility of starch and protein in sorghum, and their possible interdependence, is encapsulated in the thorough investigation by Wong et al. (2009) into the influential biochemical and structural features of sorghum grain endosperm. However, the outcomes from the Wong et al. (2009) study, and much of the relevant data in the literature, are derived from *in vitro* procedures including starch and protein digestibility assays. The extent to which data generated by *in vitro* starch and protein digestibility assays applies to the digestion of starch and protein in the gastrointestinal tract of pigs and poultry is problematic. Moreover, reports of determinations of sorghum starch, especially, and protein/amino acid digestibilities in pigs and poultry are limited; particularly when only 'tannin-free' sorghum varieties are considered.

In a recent study by Selle et al. (2013b), conventional diets based on a red sorghum (Buster) were steam-pelleted at conditioning temperatures of 65, 80 and 95°C and starch digestibility coefficients in the distal ileum ranged from 0.841 to 0.861. This is indicative of poor sorghum starch digestibility which is emphasised by the comparison of this red sorghum (Buster) with a white sorghum (Liberty) in diets steam-pelleted at 95°C where the starch digestibility of Liberty was noticeably better than Buster. The average digestibility coefficient in four sites along the small intestine favoured Liberty by 14.5% (0.852 versus 0.744) and the difference was one of 5.81% (0.910

versus 0.860) in the distal ileum. Clearly the identification of the factors in the red sorghum that contributed to its inferior starch digestibility relative to the white sorghum would be highly instructive.

The digestibility of sorghum starch is clearly inferior to that of maize, wheat and barley on the basis of *in vitro* data generated by Guiberti et al. (2012) from a total of 51 grain samples. The potential digestibility of sorghum starch (70.4%) was noticeably less than to that of maize (95.0%), wheat (92.8%) and barley (88.6%) and, interestingly, sorghum contained 65 g/kg prolamin storage proteins (or kafirin) than the other three grains, where the average was 43 g/kg on a dry matter basis.

Instructively, Taylor (2005) concluded that there is clear evidence that disulphide bond cross-linking in β- and γ-kafirin proteins is a major factor in the reduced digestibility of protein and starch in sorghum. However, it is not understood why this occurs in sorghum to a greater extent than maize, given the similarity of their prolamin proteins. Thus the enigmatic nature of sorghum as a feedstuff for broiler chickens is probably best illustrated by the lack of clarification as to the relative importance of the possible factors that are compromising starch digestibility and energy utilisation.

4.1. Kafirin and Glutelin

As stated, both starch granules and kafirin protein bodies are embedded in a glutelin protein matrix in sorghum endosperm. The close physical proximity of these three components, especially in the 'hard' (corneous/vitreous) endosperm as opposed to the 'soft' (floury/opaque) endosperm, suggests that kafirin and/or glutelin may interfere with starch digestion. The Chandrashekar and Kirleis (1988) study has already been considered in relation to pellet quality, however, this and earlier study (Chandrashekar and Desikachar, 1981), were probably the first to link protein, or more specifically kafirin, and starch gelatinisation in sorghum. A subsequent study by Zhang and Hamaker (1998) found that cooked sorghum flours had 15-25% lower *in vitro* starch digestibility than maize. However, both pepsin pretreatment and a reducing agent (sodium metabisulphite) enhanced starch digestibility prompting Zhang and Hamaker (1998) to propose that protein plays a large role in low sorghum starch digestibility. The two responsible mechanisms suggested were that (i) endosperm protein restricts starch granules from fully gelatinising or (ii) a starch-protein interaction occurs during cooking or cooling that renders gelatinised sorghum starch less digestible.

Table 5. Effect of protease supplementation on apparent digestibility coefficients of amino acids in proximal jejunum (PJ), distal jejunum (DJ), proximal ileum (PI) and distal ileum (DI) in broilers at 28 days post-hatch (adapted from Liu et al. 2013)

Site	Protease	Arg	His	Ile	Leu	Lys	Met	Phe	Thr	Val	Ala	Asp	Glu	Gly	Pro	Ser	Tyr
PJ	Nil	0.533	0.427	0.363	0.359	0.606	0.730	0.413	0.405	0.352	0.359	0.416	0.470	0.355	0.368	0.384	0.342
	Plus	0.579	0.498	0.441	0.452	0.633	0.705	0.501	0.457	0.433	0.454	0.498	0.562	0.434	0.487	0.474	0.438
DJ	Nil	0.710	0.612	0.568	0.536	0.758	0.846	0.584	0.580	0.560	0.536	0.585	0.625	0.561	0.531	0.555	0.534
	Plus	0.762	0.686	0.650	0.634	0.791	0.867	0.673	0.643	0.644	0.632	0.667	0.705	0.639	0.633	0.647	0.631
PI	Nil	0.832	0.752	0.726	0.695	0.847	0.916	0.734	0.715	0.715	0.687	0.732	0.755	0.716	0.681	0.715	0.715
	Plus	0.850	0.784	0.766	0.747	0.858	0.931	0.777	0.744	0.757	0.741	0.769	0.795	0.751	0.736	0.761	0.760
DI	Nil	0.862	0.785	0.774	0.748	0.867	0.947	0.781	0.744	0.760	0.733	0.767	0.794	0.752	0.722	0.754	0.763
	Plus	0.881	0.820	0.813	0.798	0.882	0.951	0.823	0.780	0.802	0.785	0.808	0.832	0.793	0.779	0.801	0.808

Site	Protease	Arg	His	Ile	Leu	Lys	Met	Phe	Thr	Val	Ala	Asp	Glu	Gly	Pro	Ser	Tyr
SEM		0.0233	0.0259	0.0292	0.0287	0.0221	0.0212	0.0253	0.0275	0.0310	0.0285	0.0251	0.0220	0.0287	0.025	0.0258	0.0315
Main effect: Intestinal site																	
PJ		0.556^{c}	0.462^{c}	0.402^{c}	0.405^{c}	0.619^{c}	0.717^{c}	0.457^{c}	0.431^{c}	0.393^{c}	0.407^{c}	0.457^{c}	0.516^{c}	0.394^{c}	0.428^{c}	0.429^{c}	0.390^{c}
DJ		0.736^{b}	0.649^{b}	0.609^{b}	0.585^{b}	0.774^{b}	0.857^{b}	0.628^{b}	0.611^{b}	0.602^{b}	0.584^{b}	0.626^{b}	0.665^{b}	0.600^{b}	0.582^{b}	0.601^{b}	0.582^{b}
PI		0.841^{a}	0.768^{a}	0.746^{a}	0.721^{a}	0.853^{a}	0.924^{a}	0.755^{a}	0.729^{a}	0.736^{a}	0.714^{a}	0.750^{a}	0.775^{a}	0.734^{a}	0.709^{a}	0.738^{a}	0.737^{a}
DI		0.871^{a}	0.803^{a}	0.794^{a}	0.773^{a}	0.875^{a}	0.949^{a}	0.802^{a}	0.762^{a}	0.781^{a}	0.759^{a}	0.787^{a}	0.813^{a}	0.772^{a}	0.750^{a}	0.777^{a}	0.785^{a}
Main effect: Protease																	
Nil		0.734^{b}	0.644^{b}	0.608^{b}	0.584^{b}	0.770	0.860	0.628^{b}	0.611^{b}	0.597^{b}	0.579^{b}	0.625^{b}	0.661^{b}	0.596^{b}	0.575^{b}	0.602^{b}	0.588^{b}
Plus		0.768^{a}	0.697^{a}	0.668^{a}	0.658^{a}	0.791	0.864	0.694^{a}	0.656^{a}	0.659^{a}	0.653^{a}	0.685^{a}	0.723^{a}	0.654^{a}	0.659^{a}	0.671^{a}	0.659^{a}

Table 5. (Continued)

Site	Protease	Arg	His	Ile	Leu	Lys	Met	Phe	Thr	Val	Ala	Asp	Glu	Gly	Pro	Ser	Tyr
Significance P =																	
Intestinal site		<0.001	<0.001	<0.001	<0.001	<0.001	<0.001	<0.001	<0.001	<0.001	<0.001	<0.001	<0.001	<0.001	<0.001	<0.001	<0.001
Protease		0.046	0.006	0.006	<0.001	0.174	0.782	<0.001	0.024	0.007	0.001	0.001	<0.001	0.006	<0.001	0.001	0.003
Interaction		0.832	0.770	0.806	0.755	0.946	0.715	0.639	0.919	0.839	0.776	0.683	0.518	0.804	0.492	0.691	0.722

[a,b,c] Means within columns not sharing a common superscript are significantly different ($P < 0.05$).

It is relevant that the Poultry Research Foundation has found that an inclusion of 5 g/kg sodium metabisulphite in an 'all-sorghum' broiler diet (where the grain had been steam-pelleted) significantly increased apparent ileal starch digestibility coefficients. As shown in Table 6, sodium metabisulphite increased starch digestibility by 3.76% (0.931 versus 0.966; P < 0.01) in the proximal ileum and by 3.27% (0.980 versus 0.949; P < 0.001) in the distal ileum. However, significant responses were not observed in the proximal and distal jejunum, which may indicate that the reducing agent was enhancing the digestibility of the more recalcitrant starch fractions in broilers. Nevertheless, this outcome clearly indicates that disulphide linkages in or between proteins in sorghum endosperm are depressing *in vivo* starch digestibility. One possibility is that starch granule-associated proteins are being cross-linked with kafirin and/or glutelin via disulphide bonds and their reduction enhances starch digestibility in the ileum of poultry. However, the obvious question becomes is kafirin or glutelin the more important protein fraction in this respect?

The vulnerability of sorghum protein digestibility to hydrothermal processes (eg 'moist-heating' or 'wet-cooking') is usually attributed to the formation of disulphide cross-linkages in β- and γ-kafirin in the periphery of protein bodies. Much of this data is based on *in vitro* pepsin digestibility assays; however, Selle et al. (2010c) mixed equal parts of distilled water and finely-ground sorghum for 2 hours at 45°C, which was then dried at 60 °C for 70 hours and blended into sorghum-casein broiler diets. In comparison to unprocessed, 'raw' sorghum, exposure to 'moist-heat' significantly reduced weight gain (20.0%), feed intake (21.4%), feed efficiency (4.5%) and N retention by 8.5% (0.574 versus 0.627) in broilers. The likelihood is that this hydrothermal process induced disulphide cross-linkages in sorghum proteins with very negative consequences; however, the formation of disulphide bonds is not necessarily limited to the kafirin fraction. Instructively, Emmambux and Taylor (2009) reported that wet-cooking reduced *in vitro* kafirin digestibility by 24% (0.576 versus 0.760) but wet-cooking reduced the digestibility of sorghum protein by a similar magnitude of 27%, (0.599 versus 0.826). This implies that the balance of sorghum protein, which is predominantly glutelin, is equally or even marginally more vulnerable to moist-heat than kafirin. This is not surprising in that in two sorghum varieties with an average protein content of 115 g/kg, Salunkhe et al. (1977) found that there was more cysteine in glutelin (1.39 g/kg g/100g protein) than in the kafirin (0.37 g/100g protein) fraction. However, the cystine content in kafirin is concentrated in the β- and γ-kafirin fractions (Table 4) and, conversely, Mosse et al. (1988) reported more half cystine in kafirin (0.9 g/16g N) than in the glutelin (0.55 g/16g N) fraction of sorghum.

As discussed, a *Bacillus lichenformis*-derived protease improved apparent digestibility of crude protein and amino acids in conventional, sorghum-based broiler diets (Selle et al., 2013b; Liu et al., 2013a). However, this protease also significantly increased the digestibility of starch in the distal jejunum by 13.6% (0.770 versus 0.678; $P < 0.005$) and by 4.80% (0.851 versus 0.812; $P < 0.05$) in the proximal ileum. In addition, starch digestibility was numerically enhanced in the proximal jejunum by 15.8% (0.629 versus 0.543; $P < 0.15$) and by 3.15% (0.885 versus 0.858; $P < 0.125$) in the distal ileum. Based on the presumption that this enzyme preparation did not contain any tangible starch-degrading side-activities, these findings are consistent with the proposition that kafirin and/or glutelin are interfering with starch digestion in sorghum endosperm when sorghum-based diets are offered to poultry.

There is a general acceptance of this proposition but reservations have been expressed by Gidley et al. (2011a) in respect of kafirin on the basis of scanning electron micrographs of ileal digesta in pigs offered cold-pelleted, sorghum-based diets. Gidley et al. (2011a) conclude that these micrographs are not consistent with the concept of the 'encapsulation' of sorghum starch by protein bodies. In sorghum endosperm the diameter of starch granules is considerably larger than protein bodies (~15 μm versus 1-3 μm) and the endosperm contains considerably higher concentrations of starch than kafirin. Starch granules and kafirin protein bodies are both discrete structures in sorghum endosperm. The concept that hydrothermally induced disulphide cross-linkages in the cystine-rich periphery of protein bodies are an impediment to the digestion of the central α-kafirin core is straightforward because α-kafirin would be 'protected' by a layer of poorly soluble protein. It is more difficult to conceptualise that these reactions are also a direct impediment to starch digestion, although they may physically hinder starch gelatinisation. Intuitively, it seems that any physical or chemical interactions between starch and protein that might impede starch gelatinisation/digestion are more likely to occur between starch and the glutelin protein matrix because starch granules are embedded in the protein matrix.

Another possible impediment to starch digestion that is inherent in sorghum endosperm is the endospermal cell walls. Taylor (2005) considered that it was probable that these cell walls constitute something of a barrier to digestion. The cell walls are rich in insoluble glucuronoarabinoxylan type pentosans and are intimately bound to the glutelin protein matrix, quite possibly via ferulic acid, a phenolic compound that is plentiful in sorghum.

4.2. Phytate (Myo-Inositol Hexaphosphate; IP_6)

Sorghum invariably contains phytate at absolute and relative concentrations that are often somewhat higher than other cereal grains (Doherty et al., 1982; Selle et al., 2003) and it is generally accepted that phytate negatively influences protein utilisation in poultry (Selle et al., 2000; Selle and Ravindran, 2007). This is reflected in the Ravindran et al. (1999) study where *Aspergillus niger* phytase (1200 FTU/kg) increased mean AID coefficients of 14 amino acids by 6.46% (0.791 versus 0.743) in broilers offered 'all-sorghum' diets. However, the low protein content of the sorghum (73.3 g/kg) used in this study probably indicates a relatively small kafirin proportion of total protein. There is the impression that the 'extra-phosphoric' responses to phytase in sorghum are muted in comparison to other grains (Wu et al., 2004). In another study (Selle et al., 1999), growth performance and nutrient utilisation responses to phytase in sorghum-based broiler diets were confined to diets with reduced specifications rather than standard diets. As proposed by Cosgrove (1966), at pH levels less than the isoelectric point of protein, electrostatic attractions between positively-charged arginine, histidine, and lysine residues with polyanionic phytate molecules are crucial to binary protein-phytate complex formation. If so, it follows that phytate would not readily bind kafirin due to its paucity of basic amino acids. This may in turn limit 'extra-phosphoric' responses to exogenous phytases in sorghum-based diets.

The probable negative influence of phytate on starch digestibility is not equally established but positive responses to exogenous phytase in ileal and total tract starch digestibility coefficients have been reported in poultry (Camden et al., 2001; Juanpere et al., 2005). Also, Cowieson et al. (2004b) concluded that the addition of phytase improved the digestibility of starch in the presence of IP_6 from a precision fed broiler study. Moreover, Sultan et al. (2011) reported that the addition of phytase to 'all-sorghum' broiler diets increased starch digestibility coefficients by 17.5% in the upper ileum (0.878 versus 0.840; $P < 0.05$), by 2.43% (0.927 versus 0.905; $P < 0.05$) in the lower ileum. Moreover, phytase significantly increased the rate of starch digestion by 27.4% (3.63 versus 2.85; $P < 0.05$).

In a recent experiment completed by the Poultry Research Foundation, the effects of an *Escherichia coli*-derived phytase (1000 FTU/kg) on starch digestibility coefficients in four small intestinal sites in broilers offered conventional sorghum-based diets steam-pelleted at 70°C were evaluated. Phytase numerically enhanced starch digestibility coefficients in the proximal jejunum by 25.8% (0.706 versus 0.561; $P < 0.125$), fractionally in the distal jejunum by 0.63% (0.802 versus 0.797), and by 3.50% in the proximal ileum (0.858 versus 0.829). There was a significant response of 4.64% in the distal ileum (0.880 versus 0.841; $P < 0.015$) and phytase significantly increased average starch digestibility coefficients by 7.27% (0.812 versus 0.757; $P < 0.04$) across the four small intestinal sites. Curiously, however, these positive responses were not in evidence at the higher conditioning temperatures of 80 and 90°C. This may suggest that the higher steam-pelleting conditioning temperatures modified starch and/or phytate or the nature of subsequent starch-phytate interactions in the avian gut. The higher conditioning temperatures did depress exogenous phytase activity but it remained in excess of the standard inclusion level of 500 FTU/kg at 90°C.

While not conclusive, these outcomes suggest that dietary phytate has a reciprocal, negative impact on sorghum starch digestibility given the positive responses to exogenous phytase. Lillian Thompson and her colleagues have consistently demonstrated that phytate depresses blood glycaemic indices or blood glucose responses in humans and this indicates that phytate impedes either starch digestion and/or glucose absorption (Thompson, 1988a,b). There researchers have suggested a number of mechanisms whereby phytate may depress starch digestion including phytate inhibiting α-amylase and chelating Ca, a requisite co-factor for amylase activity. There is also the possibility that phytate binds starch either directly via hydrogen bonding and/or phosphate linkages or indirectly via starch granule-associated proteins. Yoon et al. (1983) conceded that there is little information on direct interactions between phytate and starch; however, they contended that phytate is structurally capable of combining with starch via phosphate linkages. Thus direct or indirect starch-phytate interactions may impede starch digestion. As discussed by Selle at al. (2012b), it is relevant that the isoelectric point of cereal grain proteins may exceed pH 6 and the isoelectric point of kafirin is pH 5.9 (Csonka et al. 1926). Therefore it is possible that any binary complexes between starch granule-associated proteins and phytate, rather than dissociate, persist to some extent along the small intestinal tract. There is also the possibility that the formation of ternary complexes, involving Ca, could take place in the small intestine to establish an alternate link between phytate and starch granule-associated proteins.

However, there is also an indication that phytate may impede glucose absorption rather than, or in addition to, depressing starch digestibility. Demjen and Thompson (1991) reported that the addition of phytate significantly reduced glycaemic responses in humans given a carbohydrate load of glucose *per se*. It is established that phytate drags Na into the small intestinal lumen (Cowieson et al., 2004a; Selle et al., 2009) and it follows that this egress of Na would compromise glucose absorption by Na^+-dependent transport systems on which the active absorption of glucose is heavily reliant (Fuller and Reeds, 1998).

4.3. Phenolic Compounds

Sorghum contains considerably more phenolic compounds than barley, maize or wheat (Bravo, 1998). Condensed tannin, a polyphenolic compound, is a powerful anti-nutritive factor to the extent that McClymont and Duncan (1952) described it as a toxic factor and it is established that condensed tannin depresses broiler performance (Nyachoti et al., 1997). Sorghum cultivars may be divided into three categories depending on their genotypes and condensed tannin contents (Asquith et al. 1983). Type I sorghums do not have a pigmented testa and are 'tannin-free', Type II sorghums have a pigmented testa layer that contains condensed tannin and Type III ('bird-proof') sorghums contain condensed tannin in both the testa and the peripcarp. Dykes and Rooney (2006) reported condensed tannin concentrations of 0.28, 4.48 and 11.95 g/kg in Type I, Type II and Type III sorghums, respectively. It is interesting that in ostensibly 'tannin-free' Type I sorghums, Dykes and Rooney (2006) detected some, condensed tannin, albeit at very low levels.

The majority view is that current Australian sorghum crops are essentially tannin-free as they lack the pre-requisite pigmented testa (Walker, 1999; Hughes and Choct, 1999; Bryden et al. 2009b). Nevertheless, Perez-Maldonado and Rodrigues (2009) reported the presence of condensed tannin in 2004-2005 sorghums in Australia using the butanol-hydrochloric acid analytical method of Dalzell and Kerven (1998). In the 2004 harvest, the average total tannin level was 7.69 g/kg in 17 sorghums and in 2005 the average level was 5.99 g/kg in 14 sorghums. Moreover, condensed tannin concentrations in 2005-harvested sorghums were negatively correlated with live weight gain ($r = -0.723$; $P < 0.005$) and feed efficiency ($r = 0.733$; $P < 0.005$) in broilers at 21 days post-hatch (Rodrigues at al., 2007). Thus there is a clear dichotomy between the general opinion that sorghums are 'tannin-free' and the positive findings of Perez-Maldonado and Rodrigues (2009). Consequently, the eleven sorghum varieties used in the recent project are being analysed for both total phenolic compounds and condensed tannin in a quest for clarification.

According to Tomasik and Schilling (1988) phenolics readily form starch complexes and are probably more likely to form starch-phenolic complexes with amylose than amylopectin. Furthermore, Taylor et al. (2005) contended that most grain sorghum cultivars contain higher levels of phenolic compounds than other cereals and that red (non-tannin) sorghums are highly pigmented with polyphenols, anthocyanins and anthocyanidins and these phenols bind strongly to starch. This is based on the PhD thesis of Beta (1999), which concludes that phenolic compounds in sorghum influence the properties of starch. One example is ferulic acid, a phenolic acid that is plentiful in sorghum, which has been shown to influence starch pasting properties (Beta and Corke, 2004). Bjorck and Nyman (1987) found that tannic acid reduced the rate and extent of *in vitro* starch digestion by amylases and maltase. Similarly, Thompson and Yoon (1984) found that tannic acid depressed wheat starch digestion *in vitro* and Thompson et al. (1984) reported that intakes of phenolic compounds by humans were negatively correlated with blood glycaemic indices. This of course applies equally to phytate and, interestingly, as discussed by Selle et al. (2010a), phenolic compounds and phytate share several analogous, anti-nutritive properties. Phenolic compounds and phytate are located in the periphery of sorghum grain (pericarp and aleurone layer) rather than in the central endosperm where starch granules are located. Thus in native grain sorghum it seems unlikely that any inter-reactions between phenolics and phytate with starch would take place; however, they may be initiated by processing (eg hammer-milling, steam-pelleting) and may occur in the avian gut, perhaps particularly in the gizzard, which is a powerful grinding organ.

Phenolic compounds appear to be susceptible to processing and extrusion has been shown to reduce the molecular weight of phenolic polymers in sorghum (Awika et al., 2003). More recently, Dlamini et al. (2007) reported that extrusion reduced average total phenol contents by 51% (14.92 versus 7.37 mg catechin equivalents/g) in the whole grain of 5 sorghum varieties. The impact of decortication or dehulling was more profound with a reduction of 68% (14.92 versus 4.80 mg catechin equivalents/g) due to the location of phenolics in the outer layers of the grain. The effects of extrusion suggest that steam-pelleting sorghum-based diets may reduce the size of phenolic polymers, which may influence any subsequent interactions between phenolics and starch in the avian gut.

Beta et al. (2001) reported that four red sorghums had a greater concentration of total polyphenols than six white sorghums by more than a three-fold factor. Thus there is the possibility that the superior starch digestibility of the white (Liberty) over the red (Buster) sorghum observed in the Selle et al. (2013b) study may have been due to lower concentrations of phenolic compounds in the white sorghum.

4.4. Starch, Glucose and Insulin

The Poultry Research Foundation completed a project for the chicken-meat program of Rural Industries Research and Development Corporation (RIRDC) which was primarily to assess the effect of steam-pelleting temperatures of sorghum-based diets on broiler performance. However, one of the unanticipated outcomes of this project was that starch digestibility was more closely aligned to N retention than protein (N) digestibility and, superficially at least, this appears to be counter-intuitive For example, N retention was significantly correlated with average starch digestibility coefficients ($r = 0.390$; $P < 0.025$) but not average N digestibility coefficients ($r = 0.107$; $P < 0.125$) in the Selle et al. (2013b) study. However, more than 30 years ago, Fuller and Crofts (1977) considered that starch had a protein sparing effect in grower pigs when assessed as N retention. Very recently, Drew et al. (2012) demonstrated that the glycaemic index of starch influenced N retention in grower pigs. In this study grain (barley) with a low glycaemic index (71.1) supported a N retention of 0.466 but this was significantly increased by 17.4% to 0.547 by the transition to a grain (maize) with a high glycaemic index (104.8). However, N retention was not influenced by a slowly or rapidly digested protein source and their conclusion was that the rate of starch digestion is an important determinant of the efficiency of protein deposition in restricted-fed pigs.

The digestion of starch in the gut lumen permits the absorption of glucose from the small intestine which in turn prompts the pancreatic secretion of insulin and this sequence of physiological events is presumably governed by the rate and extent of starch digestion. Therefore, it is relevant that Johnston et al. (2004) reported that 500 FTU/kg phytase increased ileal starch digestibility by 2.52% (0.953 versus 0.912) in diets containing 5.0 g/kg Ca and 1.9 g/kg available P in grower pigs. Moreover, in a parallel experiment, Johnston et al. (2004) found that phytase increased plasma glucose concentrations by 5.70 % (5.19 versus 4.91 mmol/L) and insulin by 10.9% (157.2 versus 141.8 pmol/L). The increased insulin secretion in response to enhanced starch digestion and glucose absorption generated by phytase in pigs is an important finding. It probably follows that sorghum varieties with higher starch digestibility will generate higher circulating levels of glucose and insulin in the bloodstream of poultry.

There is more focus on the importance of insulin in carbohydrate and fat metabolism; nevertheless, insulin plays a complex role in protein metabolism where the liver and skeletal musculature area both targets. As reviewed by Barthel and Schmoll (2003) insulin is the most important hormone that inhibits hepatic gluconeogenesis thus sparing the deamination of gluconeogenic amino acids to provide glucose. The majority of amino acids are glucogenic as only lysine and leucine are exclusively ketogenic and are degraded to ketones rather than glucose. In skeletal muscle, insulin both increases protein synthesis and decreases protein degradation to promote a net reduction in protein turnover. For example, Kimball et al. (1997) demonstrated that insulin stimulates protein synthesis in the gastrocnemius muscle of rats and Fulks et al. (1975) reported that insulin both stimulated synthesis and inhibited protein degradation in the rat diaphragm.

Given the relationship between starch digestion, glucose absorption and insulin secretion and the positive impact of insulin on protein turnover it is not surprising that del Alamo et al. (2009) reported that the rate of starch digestion influenced performance of broilers on wheat-based diets and this should also apply to sorghum-based diets. Importantly, Tomas et al. (1998) reported that insulin-like growth factor (IGF-I) promoted lean growth and feed efficiency in broiler chickens. IGF-I significantly reduced N^r–methylhistidine excretion which is indicative of reduced rates of muscle protein breakdown and IGF-I significantly increased N retention by 13.4% from 2.24 to 2.54 g/day. Interestingly, efficiency of feed utilisation was inversely related to the rate of protein breakdown ($r^2 = 0.25$; $P < 0.01$) and it was concluded that IGF-I enhanced feed efficiency, at least in part, by attenuating rates of protein breakdown.

The digestibility of sorghum starch is relatively poor and the foregoing suggests that inferior starch digestibility compromises N retention and efficiency of feed conversion. Thus there is the need to identify the intrinsic and extrinsic factors that are responsible for the poor digestibility of sorghum starch. Extrinsic factors that appear to be involved and have been discussed include kafirin, glutelin, phytate and phenolic compounds, perhaps especially condensed tannin, if present, and ferulic acid. A range of feed additives including reducing agents, proteases, laccases (phenol-oxidising enzymes) and phytases, individually or in various combinations, may partially address the problem. More fundamentally, the identification of the important causal factors would facilitate appropriate breeding programs to select sorghum varieties with higher starch digestibility that would be a more suitable feedstuff for chicken-meat production.

5. SORGHUM GRAIN TEXTURE

The texture of grain sorghum is governed by the ratio of 'hard' (corneous/vitreous) to 'soft' (floury/opaque) fractions of the endosperm. Any consideration of sorghum texture is complicated by the various methods used to determine hardness. One method is the particle size index (PSI), which was originally developed to determine the hardness of wheat (Symes 1961, 1965). Under this method, a PSI of 8-12 is classified as "very hard" and a PSI of 20-30 as "very soft". De Alencar Figueiredo at al. (2006) determined the texture of 117 global samples of sorghum. The mean PSI was 13.6, which ranged from 8.2 to 25.7, but the dominant values were 10 and 11, which both fall into the very hard category. Recently, the Poultry Research Foundation completed a survey of 32 sorghums sourced from Queensland and New South Wales. The average PSI was 10.0 with a range of 8 to 14 which indicates that Australian sorghums have harder grain textures than sorghums grown elsewhere in the world. Also, PSI values were negatively correlated ($r = -0.438$; $P < 0.02$) with protein contents which ranged from 83 to 116 g/kg, so increasing protein, and by implication kafirin, levels were associated with increasing grain hardness. This is not surprising, several studies have shown that increasing protein and kafirin concentrations are associated with harder sorghum textures and Watterson et al. (1993) found the kafirin proportion of protein is higher in the corneous (67%) than in the floury endosperm (35%).

Cabrera (1994) determined the impact of particle size of "hard" and "soft" sorghums prior to steam-pelleting (65°C) sorghum-based diets on broiler growth performance from 7 to 28 days of age. Reductions in the mean particle size of hard sorghum from 1000 to 400 μm enhanced weight gain (5.5%), feed intake (7.7%) and feed conversion (5.1%); interestingly, the reverse was true for broilers offered soft sorghum as their best performance was observed with the coarsest particle size. Given that the majority of local sorghums are very hard the Cabrera (1994) data indicates that fine grinding to a geometric mean diameter in the order of 400 μm would be appropriate. In this context, Mikkelsen et al. (2008) reported that finely-ground (and presumably very hard) sorghum supported marked improvements in growth performance in comparison to the same coarsely-ground sorghum; whereas, in contrast, the particle size of wheat was of little consequence.

6. Particle Size and Steam-Pelleting Temperatures of Sorghum-Based Diets

Increasing attention is being paid recently to (i) the particle size of grain following grinding (usually hammer-milling) prior to its incorporation into complete diets and (ii) the conditioning temperature at which the diets are steam-pelleted in relation to broiler performance. Both aspects were investigated by the Poultry Research Foundation in the RIRDC project. However, it is generally held that the impacts of grain particle size on broiler performance are more evident in mash diets and are diminished in steam-pelleted diets (Amerah et al., 2007).

As discussed in the previous section, the optimum particle size for grain sorghum is strongly influenced by its texture or 'hardness'. Given that the eleven sorghum varieties evaluated in the project had an average PSI texture of 10.6, which is 'very hard' it was anticipated that all the sorghums would be advantaged by fine-grinding. The white sorghum (Liberty) had a PSI grain texture of 12, which was somewhat 'softer' than the average value of 10.6. This sorghum variety was evaluated in three feeding studies in which it was hammer-milled through screen sizes of 6.0, 3.2 and 2.0 mm prior to its incorporation into conventional diets that were steam-pelleted at 90-95°C. The comparative results are shown in Table 7 and should be interpreted with some caution as they are derived from three separate, but very similar, feeding studies. The reduction in screen size from 6.0 to 3.2 mm improved weight gain by 5.35%, and feed conversion by 7.99%; this reduction also enhanced the majority of the parameters assessed. However, the transition from 3.2 to a 2.0 mm screen size depressed weight gain by 0.71% and feed conversion by 4.85% and digestibility coefficients of starch and nitrogen were depressed by, on average, 4.93% (0.810 versus 0.810) and 4.27% (0.672 versus 0.702), respectively. Efficiency of feed conversion is a critical parameter for profitable chicken-meat production. Therefore, it is noteworthy that there is a significant quadratic relationship ($r = 0.787$; $P < 0.05$) between hammer-mill screen size and FCR as illustrated in Figure 4. The relevant quadratic equation is a follows:

It may be deduced from this equation that for this particular sorghum, with a PSI texture of 12, that a hammer-mill screen size of 3.67 mm would be optimal generating a feed conversion ratio of 1.485. This clear-cut influence of sorghum particle size as dictated by hammer-mill sieve size on feed conversion efficiency was not expected. This outcome suggests that more attention should be paid to the impacts of hammer-milling sorghum through differing sieve sizes on geometric mean and distribution of grain particle size prior to its incorporation into steam-pelleted diets.

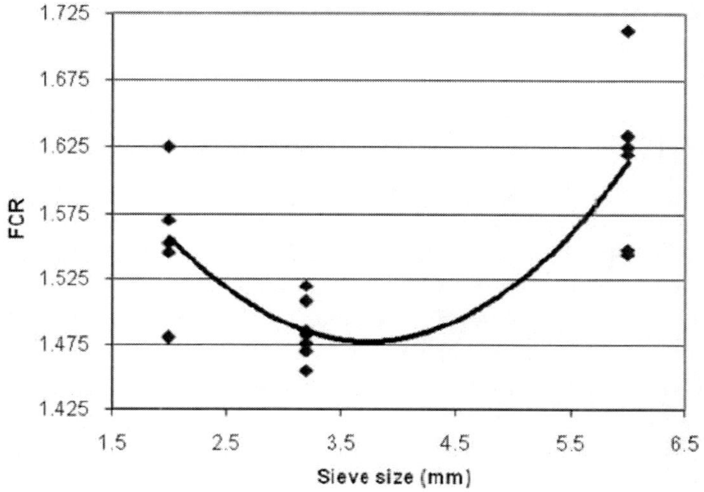

Figure 4. The quadratic relationship ($r = 0.787$; $P < 0.05$) between screen size for hammer-milling white sorghum (Liberty) prior to incorporation into diets steam-pelleted at 90-95°C and feed conversion ratios of broiler chickens.

$$y_{(FCR)} = 1.848 - 0.198*\text{sieve size(mm)} + 0.027*[\text{sieve size(mm)}]^2.$$

The RIRDC project consisted of six feeding studies in which eleven sorghum varieties (9 red, 1 white, 1 yellow) were steam-pelleted at conditioning temperatures ranging from 65 to 97°C. A key finding was that increasing conditioning temperatures linearly reduced Promatest protein solubility of the diets and this reduction was more profound in sorghum than the balance of dietary protein sources. Moreover, about 30% of this reduction in protein solubility may be attributed to the formation of disulphide cross-linkages between cysteine residues as demonstrated by decreases in concentrations of free sulphydryl groups and increases in concentrations of disulphide bonds following steam-pelleting, which were correlated to Promatest protein solubility. The likelihood is the majority of these disulphide cross-linkages induced by steam-pelleting occurred in the glutelin protein matrix and β- and γ-kafirin fractions in the periphery of protein bodies in sorghum endosperm.

It is instructive to compare the effects of steam-pelleting diets based on two red sorghums at conditioning temperatures of 65 and 97ºC in the Selle et al. (2013a) study when offered only as intact pellets. The higher conditioning temperature increased feed intake by 3.31% but weight gain by only 0.68% so that feed conversion ratios were inferior by 2.46% (1.549 versus 1.588). Taken individually, the higher conditioning temperature depressed the FCR of the first sorghum by 2.54% (1.535 versus 1.497; P = 0.058) and that of the second sorghum by 2.56% (1.641 versus 1.600; P = 0.041). The inferior feed efficiency at the higher conditioning temperature was associated with relatively modest decreases in AME, N retention, AMEn and ileal N digestibility coefficients. However, there was a substantial 8.69% reduction in jejunal N digestibility coefficients from 0.518 to 0.473 where the higher conditioning temperature reduced the protein solubilities of the two diets from an average of 44.4 to 33.5%. That increasing the conditioning temperature from 65 to 97ºC reduced feed efficiency by nearly 2.5% is a relatively subtle but pivotal outcome because of the importance of efficient feed conversion in chicken-meat production.

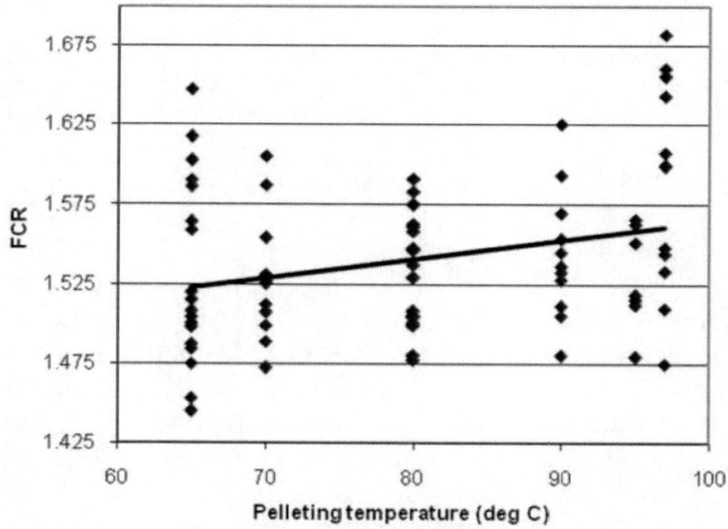

Figure 5. The linear relationship (r = 0.284; P < 0.015) between conditioning temperatures at which sorghum-based diets were steam-pelleted and feed conversion ratios of broiler chickens.

Table 6. The effect of sodium metabisulphite (5.0 g/kg) on apparent starch digestibility coefficients in the proximal and distal jejunum, proximal and distal ileum of broiler chickens offered 'all-sorghum' diets

Treatment Sodium metabisulphite	Proximal jejunum	Distal jejunum	Proximal ileum	Distal ileum
0 g/kg	0.732	0.880	0.931[a]	0.949[a]
5.0 g/kg	0.696	0.876	0.966[b]	0.980[b]
SEM	0.0143	0.0154	0.0071	0.0047
Significance (P =)	0.115	0.844	0.007	< 0.001
LSD (P < 0.05)	-	-	0.0227	0.0149

Table 7. The impact of hammer-milling white sorghum #8 (Liberty) through different screen sizes prior to incorporation into diets steam-pelleted at conditioning temperatures of 90-95°C on broiler performance

Screen size Conditioning temperature	6.0 mm 95°C	3.2 mm 95°C	2.0 mm 90°C
Item			
Weight gain (g/bird)	1462	1540	1529
Feed intake (g/bird)	2359	2285	2382
Feed conversion ratio	1.614	1.485	1.557
Relative gizzard weight (g/kg)	19.35	17.79	16.90
AME (MJ/kg DM)	13.39	12.79	13.31
N retention (%)	56.8	60.3	64.5
AMEn (MJ/kg DM)	11.25	11.80	11.68
Starch digestibility coefficients			
Proximal jejunum	0.606	0.759	0.721
Distal jejunum	-	0.838	0.783
Proximal ileum	0.844	0.902	0.847
Distal ileum	0.869	0.910	0.889
Nitrogen digestibility coefficients			
Proximal jejunum	0.488	0.540	0.522
Distal jejunum	-	0.681	0.642
Proximal ileum	0.844	0.793	0.741
Distal ileum	0.869	0.795	0.781

The effects of different conditioning temperatures of diets fed as intact pellets were also assessed in the third, fifth and sixth feeding studies of the project. As shown in Figure 5, there is a significant linear relationship ($r = 0.284$; $P < 0.015$) between steam-pelleting temperatures and feed conversion ratios of broiler chickens. The correlation is not strong, which is understandable given the variations in sorghum variety and grain particle size across the three feeding studies but the significant relationship indicates that increasing conditioning temperatures of sorghum-based broiler diets compromise efficiency of feed conversion.

Presently, 'whole grain feeding', where whole grain may represent 20% of the diet with the balance made up by a pelleted concentrate is a well accepted practice in Australia. The relevant scientific literature indicates that whole grain feeding noticeably increases relative gizzard weights but this appears to be associated with improvements in feed conversion efficiency, which presumably stems from enhanced gizzard function (Cowieson and Selle, 2011). Across the project, both increases in conditioning temperatures ($r = 0.431$; $P < 0.10$) and sorghum particle size ($r = 0.436$; $P < 0.10$) tended to be associated with heavier relative gizzard weights. However, taken together, there is a significant relationship ($r = 0.596$; $P < 0.05$) between conditioning temperature (ºC) and hammer-mill screen size (mm) where the multiple linear regression equation is as follows:

$$y_{(\text{relative gizzard weight, g/kg})} = 12.367 + 0.06063{*}^{\circ}\text{C} + 0.470{*}\text{mm}.$$

In essence, higher conditioning temperatures for steam-pelleting sorghum-based broiler diets have a negative impact on efficiency of feed conversion, protein solubility and N digestibility coefficients. However, the negative impact is moderate and this is probably due to two compensatory factors. High steam-pelleting temperatures are associated with better quality and harder textured pellets, where, importantly, the merits of pellets with higher breaking forces on broiler performance have been demonstrated (Parsons et al., 2006). It may be possible to produce hard pellets at low temperatures with the use of binding agents but this approach needs to be viable in practice. In addition, increasing conditioning temperatures at which mediumly-ground sorghum-based diets were steam-pelleted from 65 to 95-97 ºC significantly increased relative gizzard weights by 8.1% in one feeding study (20.79 versus 19.24 g/kg) and by 8.4% (19.38 versus 17.88 g/kg) in a second study (Selle et al., 2013a,b). Thus the negative impacts of high conditioning temperatures on protein solubility are probably attenuated by compensatory increases in gizzard weight, and presumably function. Thus high steam-pelleting temperatures of sorghum-based diets can be expected to increase relative gizzard weights and this is probably to the benefit of feed efficiency and also starch and N digestibility from reductions in feed intake rates. Thus on one hand high steam-pelleting temperatures may decrease N digestibility by reducing protein solubility but, on the other, tend to increase N digestibility by stimulating gizzard function.

CONCLUSION

It is our contention that the performance of broiler chickens offered sorghum-based diets is not necessarily inferior to wheat and may in fact be fully comparable. However, we concede that inconsistency is an issue where wheat-based diets are advantaged by NSP-degrading feed enzymes with predominantly xylanase activity.

Nevertheless, strategies can be recommended to enhance the performance of broiler chickens offered sorghum-based diets. A basic issue is the challenge of adequate pellet quality. This can be addressed in the feed-mill, at least to a partial extent, as anecdotal evidence suggests that certain local integrated feed-mills have developed processing strategies whereby sorghum-based diets with acceptable pellet quality are produced. It is quite possible that this would be facilitated by the dietary inclusion of reducing agents on the basis of our initial investigation. It also seems likely that the selection of 'low-kafirin' sorghums in breeding programs would enhance pellet quality. Clearly, the conditioning temperatures at which sorghum-based diets are steam-pelleted should be moderated without, ideally, compromising pellet quality. Certainly, more attention should be paid to sorghum particle size following hammer-milling as grain particle size, even in steam-pelleted diets, appears to hold considerable importance. However, the optimal grain particle size will be dependent on both grain textures and conditioning temperatures.

The problem presented by the poor protein quality of sorghum is confined by the fact that sorghum provides about 30% of total dietary protein but it still can be addressed. There are tangible variations in both concentration and digestibility of amino acids; consequently, it is questionable if intended dietary amino acid specifications are being met accurately in the formulation of sorghum-based diets when amino acid concentrations are derived from sorghum protein contents. However, this challenge can be overcome by determining the amino acid profile of incoming sorghum shipments by NIR spectroscopy. Then it is possible to formulate sorghum-based diets on the basis of established sorghum amino acid levels. Additionally, when sorghum-based diets are being formulated on the basis of ileal digestible rather than dietary amino acids it would be prudent to apply conservative ileal digestible values. Also, the dietary inclusion costs of synthetic lysine and methionine permit the fortification of sorghum-base diets with these two critical amino acids. Finally, the *Bacillus lichenformis*-derived protease holds promise as a means of enhancing the quality of protein/amino acids in sorghum and this may also apply to reducing agents such as sodium metabisulphite.

In our view, the major reason that sorghum is an enigmatic grain for chicken-meat production is the poor utilisation of energy derived from starch coupled with the fact that the underlying causal factors remain unclarified. Kafirin is frequently implicated as perhaps the single most important negative factor. The authors are sceptical that this is the case but, if established, the desirability of selecting 'low-kafirin' sorghums becomes especially real. Perhaps glutelin is a more important negative factor than kafirin in terms of starch digestibility. However, to the extent that proteins in sorghum endosperm are limiting starch digestibility, the dietary inclusions of a particular protease and a reducing agent such as sodium metabisulphite appear to hold promise. Indeed, their combined inclusions could hold particular promise given that it is unlikely that the protease has the capacity to cleave disulphide bonds; whereas, sodium metabisulphite has this capacity.

However, it is our suggestion that both phenolic compounds and phytate, perhaps in tandem, are negatively influencing sorghum starch digestibility. This could be clarified by comparing rates of starch digestion and starch digestibility coefficients between whole and dehulled sorghum grain as the dehulling procedure would eliminate the majority of phenolics and phytate due to their peripheral location in the aleurone layer and pericarp of grain sorghum. Moreover, it may be feasible to assess the relative importance of glutelin and kafirin on starch digestibility in dehulled sorghums. Broiler diets based on a low protein sorghum, a high protein sorghum and an equal blend should provide a range of kafirin:glutelin ratios, which would permit the determination of their impacts on starch digestibility coefficients and starch digestion rates. Finally, it is our contention, that sorghum would cease being an enigmatic grain for chicken-meat production if the reasons for the poor utilisation of energy derived from starch are elucidated so that they can be addressed.

ACKNOWLEDGMENTS

The financial support provided by the Chicken-meat Program of Rural Industries Research and Development Corporation (RIRDC) to the Poultry Research Foundation for the completion of grain sorghum-related projects is gratefully acknowledged.

REFERENCES

ACMF (2011). The Australian Chicken Meat Industry: an industry in profile. Australian Chicken Meat Federation (ACMF) Inc. North Sydney, NSW.

Amerah, A. M., Ravindran, V., Lentle, R. G. & Thomas, D. G. (2007). Feed particle size: Implications on the digestion and performance of poultry. *World's Poultry Science Journal*, *63*, 439-455.

Asquith, T. N., Izuno, C. C. & Butler, L. G. (1983). Characterization of the condensed tannin (proanthocyanidin) from a group II sorghum. *Journal of Agricultural and Food Chemistry*, *31*, 1299-1303.

Awika, J. M., Dykes, L., Gu, L., Rooney, L. W. & Prior, R. L. (2003). Processing of sorghum (*Sorghum bicolor*) and sorghum products alters procyanidin oligomer and polymer distribution and content. *Journal of Agricultural and Food Chemistry*, *51*, 5516-5521.

Barthel, A. & Schmoll, D. (2003). Novel concepts in insulin regulation of hepatic gluconeogenesis. *American Journal of Physiology (Endocrinology and Metabolism)*, *285*, E685-E692.

Beckwith, A. C. (1972). Grain sorghum glutelin: isolation and characterization. *Journal of Agricultural and Food Chemistry*, *20*, 761-764.

Beta, T. (1999). Processing of polyphenols-rich sorghums. PhD Thesis. University of Pretoria. Pretoria, RSA.

Beta, T. & Corke, H. (2001). Genetic and environmental variation in sorghum starch properties. *Journal of Cereal Science*, *34*, 261-268.

Beta, T. & Corke, H. (2004). Effect of ferulic acid and catechin on sorghum and maize starch pasting properties. *Cereal Chemistry*, *81*, 418-422.

Bjork, I. M. & Nyman, M. E. (1987). *In vitro* effects of phytic acid and polyphenols on starch digestion and fiber degradation. *Journal of Food Science*, *52*, 1588-1594.

Black, J. L., Hughes, R. J., Nielsen, S. G., Tredrea, A. M., MacAlpine, R. & van Barneveld, R. J. (2005). The energy value of cereal grains, particularly wheat and sorghum, for poultry. *Proceedings, Australian Poultry Science Symposium, 17*, 21-29.

Bravo, L. (1998). Polyphenols: Chemistry, dietary sources, metabolism, and nutritional significance. *Nutrition Reviews, 56*, 317-333.

Briggs, J. L., Maier, D. E., Watkins, B. A. & Behnke, K. C. (1999). Effect of ingredients and processing parameters on pellet quality. *Poultry Science, 78*, 1464-1471.

Bryden, W. L., Li, X., Ravindran, G., Hew, L. I. & Ravindran, V. (2009). Ileal digestible amino acids in feedstuffs for poultry. RIRDC Publication 09/071. Rural Industries Research and Development Corporation, Barton, ACT.

Bryden, W. L., Selle, P. H., Cadogan, D. J., Li, X., Muller, N. D., Jordan, D. R., Gidley, M. J. & Hamilton, W. D. (2009b). A review of the nutritive value of sorghum for broilers. RIRDC Publication 09/077. Rural Industries Research and Development Corporation, Barton, ACT.

Cabrera, M. R. (1994). Effects of sorghum genotype and particle size on milling characteristics and performance of finishing pigs, broiler chicks, and laying hens. MSc Thesis. Kansas State University. Manhattan, KS.

Camden, B. J., Morel, P. C. H., Thomas, D. V., Ravindran, V. & Bedford, M. R. (2001). Effectiveness of exogenous microbial phytase in improving the bioavailabilities of phosphorus and other nutrients in maize-soya-bean meal diets for broilers. *Animal Science, 73*, 289-297.

Chandrashekar, A. & Desikachar, H. S. R. (1981). Studies on the hydration of starches, flour and semolina from different cereal grains. *Journal of Food Science and Technology, 18*, 12-??

Chandrashekar, A. & Kirleis, A. W. (1988). Influence of protein on starch gelatinization in sorghum. *Cereal Chemistry, 65*, 457-462.

Cosgrove, D. J. (1966). The chemistry and biochemistry of inositol polyphosphates. *Reviews of Pure and Applied Chemistry, 16*, 209-224.

Cowieson, A. J., Acamovic, T. & Bedford, M. R. (2004a). The effects of phytase and phytic acid on the loss of endogenous amino acids and minerals from broiler chickens. *British Poultry Science, 45*, 101-108.

Cowieson, A. J., Acamovic, T. & Bedford, M. R. (2004b). The effect of phytic acid and phytase on the digestibility of maize starch for growing chickens. *Proceedings of the International Poultry Scientific Forum* p31 Atlanta GA.

Cowieson, A. J. & Selle, P. H. (2011). The environmental impact of low feed conversion ratios in poultry. *Recent Advances in Animal Nutrition – Australia*, *18*, 157-164.

Csonka, F. A., Murphy, J. C. & Jones, D. B. (1926). The iso-electric points of various proteins. *Journal of the American Chemical Society*, *48*, 763-768.

Dalzell, S. A. & Kerven, G. L. (1998). A rapid method for the measurement of *Leucaena* spp proanthocyanidins by the proanthocyanidin (butanol/HCl) assay. *Journal of the Science in Food and Agriculture*, *78*, 405-416.

De Alencar Figeiredo, L. F., Davrieux, F., Fliedel, G., Rami, J. F., Chantreau, J., Deu, M., Cortois, B. & Mestres, C. (2006). Development of NIRS equations for food grain quality traits through exploitation of a core collection of cultivated sorghum. *Journal of Agricultural and Food Chemistry*, *54*, 8501-8509.

De Mesa-Stonestreet, N. J., Alavi, S. & Bean, S. R. (2010). Sorghum proteins: the concentration, isolation, modification, and food application of kafirins. *Journal of Food Science*, *75*, R90-R104.

Demjen, A. P. & Thompson, L. U. (1991). Calcium and phytic acid independently lower the glycemic response to a glucose load. *Proceedings of the 34th Canadian Federaton of Biological Sciences* p53 (Abstr).

Dlamini, N. R., Taylor, J. R. N. & Rooney, L. W. (2007). The effect of sorghum type and processing on the antioxidant properties of African sorghum-based foods. *Food Chemistry*, *105*, 1412-1419.

Doherty, C., Faubion, J. M. & Rooney, L. W. (1982). Semiautomated determination of phytate in sorghum and sorghum products. *Cereal Chemistry*, *59*, 373-377.

Drew, M. D., Schafer, T. C. & Zijlstra, R. T. (2012). Glycemic index of starch affects nitrogen retention in grower pigs. *Journal of Animal Science*, *90*, 1233-1241.

Duodu, K. G., Taylor, J. R. N., Belton, P. S. & Haymaker, B. R. (2003). Factors affecting sorghum protein digestibility. *Journal of Cereal Science*, *38*, 117-131.

Dykes, L. & Rooney, L. W. (2006). Review. Sorghum and millet phenols and antioxidants. *Journal of Cereal Science*, *44*, 236-251.

Emmambux, N. M. & Taylor, J. R. N. (2009). Properties of heat-treated sorghum and maize meal and their prolamin proteins. *Journal of Agricultural and Food Chemistry*, *57*, 1045-1050.

Fuller, M. F. & Crofts, R. M. J. (1977). The protein-sparing effect of carbohydrate 1. Nitrogen retention of growing pigs in relation to diet. *British Journal of Nutrition*, *38*, 479-488.

Fuller, M. F. & and Reeds, P. J. (1998). Nitrogen cycling in the gut. *Annual Reviews in Nutrition*, *18*, 385-411.

Fulks, R. M., Li, J. B. & Goldberg, A. L. (1975). Effects if insulin, glucose, and amino acids on protein turnover in the rat diaphragm. *Journal of Biological Chemistry*, *250*, 290-298.

Gidley, M. J., Flanagan, B. M., Sharpe, K. & Sopade, P. A. (2011a). Starch digestion in monogastrics – mechanisms and opportunities. *Recent Advances in Animal Nutrition – Australia*, *18*, 207-213. University of New England. Armidale, NSW.

Gidley, M. J., Pluschke, A. M., Sopade, P. A., Al-Rabadi, G. J. S., Sultan, A., Gan, C. Y., Li, X., Zhang, D. & Bryden, W. L. (2011b). Sorghum grain starch digestibility: effects of particle size and enzyme treatment. *Proceedings, Australian Poultry Science Symposium*, *22*, 139-146.

Giuberti, G., Gallo, A., Cerioli, C. & Masoera, F. (2012). *In vitro* starch digestion and predicted glycaemic index of ceral grains commonly utilized in pig nutrition. *Animal Feed Science and Technology*, *174*, 163-173.

Hamaker, B. R., Mohamed, A. A., Habben, J. E., Huang, C. P. & Larkins, B. A. (1995). Efficient procedure for extracting maize and sorghum kernel proteins reveals higher prolamin contents than conventional methods. *Cereal Chemistry*, *72*, 583-588.

Hughes, R. J. & Choct, M. (1999). Chemical and physical characteristics of grains related to variability in energy and amino acid availability in poultry. *Australian Journal of Agricultural Research*, *50*, 689-701.

Hughes, R. J. & Choct, M. (1999). Chemical and physical characteristics of grains related to variability in energy and amino acid availability in poultry. *Australian Journal of Agricultural Research*, *50*, 689-701.

Johnston, S. L., Williams, S. B., Southern, L. L., Bidner, T. D., Bunting, L. D., Matthews, J. O. & Olcott, B. M. (2004). Effect of phytase addition and dietary calcium and phosphorus levels on plasma metabolites and ileal and total tract nutrient digestibility in pigs. *Journal of Animal Science*, *82*, 705-714.

Juanpere, J., Perez-Vendrell, A. M. & Brufau, J. (2004). Effect of microbial phytase on broilers fed barley-based diets in the presence or not of endogenous phytase. *Animal Feed Science and Technology*, *115*, 265-279.

Kimball, S. R., Jurasinski, C. V., Lawrence, J. C. & Jefferson, L. S. (1997). Insulin stimulates protein synthesis in skeletal muscle by enhancing the association of eIF-4E and eIF-4G. *American Journal of Physiology (Cellular Physiology)*, *272*, C754-C759.

Landry, J. & Moureaux, T. (1970). Héterogénéité des glutélines du grain de mais: extraction sélective et composition en acides aminés des trios fractions isolées. [Heterogeneity of glutelins of corn – selective extraction of amino acids of three isolated fractions] *Bulletin de la Societe de Chimique Biologique*, *52*, 1021-1037.

Liu, S. Y., Selle, P. H., Court, S. G. & Cowieson, A. J. (2013a). Protease supplementation enhances apparent digestibilities of amino acids at four small intestinal sites in broiler chickens offered sorghum-based diets. *Proceedings*, *Australian Poultry Science Symposium*, *24*, 41-44.

Liu, S. Y., Selle, P. H., Gill, R. J. & Cowieson, A. J. (2013b). Starch and nitrogen digestion kinetics influence growth performance and nitrogen retention in red sorghum–based broiler diets. *Proceedings*, *Australian Poultry Science Symposium*, *24*, 115-118.

McClymont, G. L. & Duncan, D. C. (1952). Studies on nutrition of poultry. III. Toxicity of grain sorghum for chicks. *Australian Veterinary Journal*, *28*, 229-233.

Mikkelsen, L. L., Yan, S., Goopy, J. P. & Ili, P. A. (2008). Effect of grain type and particle size on growth performance and intestinal microbial populations in broiler chickens. *Proceedings*, *XXIII World's Poultry Congress*. World's Poultry Science Association. 30th June to 4th July 2008, Brisbane Qld.

Morrison, C. D., Xi, X., White, C. L., Ye, J. & Martin, R. J. (2007). Amino acids inhibit *Agrp* gene expression via an mTOR-dependent mechanism. *American Journal of Physiology - Endocrinology and Metabolism*, *293*, E165-E171.

Mosse, J., Huet, J. C. & Baudet, J. (1988). The amino acid composition of whole sorghum grain in relation to its nitrogen content. *Cereal Chemistry*, *65*, 271-277.

Nyachoti, C. M., Atkinson, J. L. & Leeson, S. (1997). Sorghum tannins: a review. *World's Poultry Science Journal*, *53*, 5-21.

Perez-Maldonado, R. A. & Rodrigues, H. D. (2009). Nutritional characteristics of sorghums from Queensland and New South Wales for Chicken meat production. RIRDC Publication No 09/170. Rural Industries Research and Development Corporation. Barton, ACT.

Ravindran, V., Hew, L. I. & Bryden, W. L. (1998). Digestible amino acids in poultry feedstuffs. RIRDC Publication No. 98/9. Rural Industries Research and Development Corporation, Barton, ACT.

Ravindran, V., Cabahug, S., Ravindran, G. & Bryden, W. L. (1999). Influence of microbial phytase on apparent ileal amino acid digestibility in feedstuffs for broilers. *Poultry Science*, *78*, 699-706.

Rodrigues, H. D., Perez-Maldonado, R. A., Trappett, P., Barram, K. M. & Kemsley, M. (2007). Broiler performance in Australian sorghum-based starter and finisher diets (2005 harvest). *Proceedings, Australian Poultry Science Symposium*, *19*, 93-96.

Salunkhe, D. K., Kadam, S. S. & Chavan, J. K. (1977). Nutritional quality of proteins in grain sorghum. *Qualitas Plantarum – Plant Foods for Human Nutrition*, *27*, 187-205.

Seckinger, H. L. & Wolf, J. (1973). Sorghum protein ultrastructures as it relates to composition. *Cereal Chemistry*, *50*, 455-464.

Selle, P. H., Ravindran, V., Pittolo, P. H. & Bryden, W. L. (1999). An evaluation of microbial phytase in sorghum-based broiler diets. *Proceedings, Australian Poultry Science Symposium*, *11*, 97-100.

Selle, P. H., Ravindran, V., Caldwell, R. A. & Bryden, W. L. (2000). Phytate and phytase: Consequences for protein utilisation. *Nutrition Research Reviews*, *13*, 255-278.

Selle, P. H., Walker, A. R. & Bryden, W. L. (2003). Total and phytate-phosphorus contents and phytase activity of Australian-sourced feed ingredients for pigs and poultry. *Australian Journal of Experimental Agriculture*, *45*, 475-479.

Selle, P. H. & Ravindran, V. (2007). Microbial phytase in poultry nutrition. *Animal Feed Science and Technology*, *135*, 1-41.

Selle, P. H., Partridge, G. G. & Ravindran, V. (2009). Beneficial effects of xylanase and/or phytase inclusions on ileal amino acid digestibility energy utilisation mineral retention and growth performance in wheat-based broiler diets. *Animal Feed Science and Technology*, *153*, 303-313.

Selle, P. H., Cadogan, D. J., Li X. & Bryden, W. L. (2010a). Implications of sorghum in broiler chicken nutrition. *Animal Feed Science and Technology*, *156*, 57-74.

Selle, P. H., Cadogan, D. J., Ru, Y. J. & Partridge, G. G. (2010b). Impact of exogenous enzymes in sorghum- or wheat-based broiler diets on nutrient utilization and growth performance. *International Journal of Poultry Science*, *9*, 53-58.

Selle, P. H., Gill, R. J. & Downing, J. A. (2010c). The vulnerability of sorghum to 'moist-heat'. *Proceedings, Australian Poultry Science Symposium*, *21*, 68-71.

Selle, P. H. (2011). The protein quality of sorghum. *Proceedings, Australian Poultry Science Symposium*, *22*, 147-160.

Selle, P. H., Liu, S. Y., Cai, J. & Cowieson, A. J. (2012a). Steam-pelleting and feed form of broiler diets based on three coarsely ground sorghums influences growth performance, nutrient utilisation, starch and nitrogen digestibility. *Animal Production Science*, *52*, 842-852.

Selle, P. H., Cowieson, A. J., Cowieson N. P. & Ravindran, V. (2012). Protein-phytate interactions in pig and poultry nutrition., a reappraisal. *Nutrition Research Reviews*, *25*, 1-17.

Selle, P. H., Gill, R. J. & Cowieson, A. J. (2013a). Steam-pelleting temperatures, sorghum variety and feed form influence growth performance, nutrient utilisation and nitrogen digestibility in broiler chickens. *Proceedings, Australian Poultry Science Symposium*, *24*, 142-145.

Selle, P. H., Liu, S. Y., Cai, J. & Cowieson, A. J. (2013b). Steam-pelleting temperatures, grain variety feed form and protease supplementation of mediumly-ground, sorghum-based broiler diets. I. Influence on growth performance, relative gizzard weights, nutrient utilisation, starch and nitrogen digestibility. *Animal Production Science*, *53*, (accepted for publication).

Shull, J. M., Watterson, J. J. & Kirleis, A. W. (1992). Purification and immunocytochemical localization of kafirins in *Sorghum bicolor* (L. Moench) endosperm. *Protoplasma*, *171*, 64-74.

Sultan, A., Li, X., Cadogan, D. J. & Bryden, W. L. (2010). Dietary enzymes alter sorghum protein digestibility and AME content. *Proceedings, Australian Poultry Science Symposium*, *21*, 94.

Sultan, A., Gan, C. Y., Li, X., Zhang, D. & Bryden, W. L. (2011a). Dietary enzyme combinations improve sorghum ileal protein an starch digestibility during the broiler starter phase. *Proceedings, Australian Poultry Science Symposium*, *22*, 82.

Sultan, A., Gan, C. Y., Li, X., Zhang, D. & Bryden, W. L. (2011b). Dietary enzymes modulate sorghum starch digestion kinetics in broilers. *Proceedings, Australian Poultry Science Symposium*, *22*, 83.

Sultan, A., Li, X., Zhang, D. & Bryden, W. L. (2011c). Feed enzymes improve ileal protein, starch and mineral digestibility of sorghum by broilers. *Recent Advances in Animal Nutrition – Australia*, *18*, 143-144.

Symes, K. J. (1961). Classification of Australian wheat varieties based on the granularity of their wholemeal. *Australian Journal of Experimental Agriculture and Animal Husbandry*, *1*, 18-23.

Symes, K. J. (1965). The inheritance of grain hardness in wheat as measured by the particle size index. *Australian Journal of Agricultural Research, 16*, 113-123.

Taylor, J. R. N., Schüssler, L. & van der Walt, W. H. (1984). Fractionation of proteins from low-tannin sorghum grain. *Journal of Agricultural and Food Chemistry, 32*, 149-154.

Taylor, J. R. N. & Dewar, J. (2001). Developments in sorghum food technologies. *Advances in Food and Nutrition Research, 43*, 217-264.

Taylor, J. R. N. (2005). Non-starch polysaccharides, protein and starch: form function and feed – highlights on sorghum. *Proceedings, Australian Poultry Science Symposium, 17*, 9-16.

Thompson, L. U. & Yoon, J. H. (1984). Starch digestibility as affected by polyphenols and phytic acid. *Journal of Food Science, 49*, 1228-1229.

Thompson, L. U., Yoon, J. H., Jenkins, D. J. A., Wolever, T. M. S. & Jenkins, A. L. (1984). Relationships between polyphenol intake and blood glucose response on normal and diabetic individuals. *The American Journal of Clinical Nutrition, 39*, 745-751

Thompson, L. U. (1988a). Antinutrients and blood glucose. *Food Technology, 42*, 123-131.

Thompson, L. U. (1988b). Phytic acid: a factor influencing starch digestibility and blood glucose response. In *Phytic Acid: Chemistry and Applications* [E Graf, editor] pp 173-194. Pilatus Press, Minneapolis.

Tomas, F. M., Pym, R. A., McMurtry, J. P. & Francis, G. L. (1998). Insulin-like growth factor IGF-I but not IGF-II promotes lean growth and feed efficiency in broiler chickens. *General and Comparative Endocrinology, 110*, 262-275.

Tomasik, P. & Schilling, C. H. (1988). Complexes of starch with organic guests. *Advances in Carbohydrate Chemistry and Biochemistry, 53*, 345-426.

Virupaksha, T. K. & Sastry, L. V. S. (1968). Studies on the protein content and amino acid composition of some varieties of grain sorghum. *Journal of Agricultural and Food Chemistry, 16*, 199-203.

Walker, T. (1999). Sorghum grain for poultry and swine. *ASA Technical Bulletin*, AN20-1999, 1-10.

Wallace, J. C., Lopes, M. A., Lopes, A., Paiva, E. & Larkins, B. A. (1990). New methods for extraction and quantitation of zeins reveal a high content of γ-zein in modified opaque-2 maize. *Plant Physiology, 92*, 191-196.

Watterson, J. J., Shull, J. M., Mohamed, A. M. A., Reddy, V. & Kirleis, A. W. (1990). Isolation of high-cysteine kafirin protein and its cross-reactivity with gamma-zein antiserum. *Journal of Cereal Science*, *12*, 137-144.

Watterson, J. J., Shull, J. M. & Kirleis, A. W. (1993). Quantitation of α-, β-, and γ-kafirins in vitreous and opaque endosperm of *Sorghum bicolour*. *Cereal Chemistry*, *70*, 452-457.

Wong, J. H., Lau, T., Cai, N., Singh, J., Pedersen, J. F., Vensel, W. H., Hurkman, W. J., Wilson, J. D., Lemaux, P. G. & Buchanan, B. B. (2009). Digestibility of protein and starch from sorghum (*Sorghum bicolour*) is linked to biochemical and structural features of grain endosperm. *Journal of Cereal Science*, *49*, 73-82.

Wood, J. F. (1987). The functional properties of feed raw materials and their effect on the production and quality of feed pellets. *Animal Feed Science and Technology*, *18*, 1-17.

Wu, Y. B., Ravindran, V. & Hendriks, W. H. (2004). Influence of exogenous enzyme supplementation on energy utilisation and nutrient digestibility of cereals for broilers. *Journal of the Science of Food and Agriculture*, *84*, 1817-1822.

Yoon, J. H., Thompson, L. U. & Jenkins, D. J. A. (1983). The effect of phytic acid on in vitro rate of starch digestibility and blood glucose response. *The American Journal of Clinical Nutrition*, *38*, 835-852.

Zhang, G. & Hamaker, B. R. (1998). Low α-amylase starch digestibility of cooked sorghum flours and the effect of protein. *Cereal Chemistry*, *75*, 710-713.

In: Sorghum
Editor: Patricia C. Parra
ISBN: 978-1-62618-351-3

Chapter 2

UNLOCKING THE POTENTIAL OF SORGHUM FOR DEVELOPMENT IN EAST AFRICA

Tom Martin[1], Moses Biruma[2], Johan Fogelqvist[1], Patrick Okori[3] and Christina Dixelius[1*]

[1]Dept. Plant Biology and Forest Genetics, Uppsala Biocenter, Swedish University of Agricultural Sciences & Linnean Center for Plant Biology, Uppsala, Sweden

[2]National Agriculture Research Organization, Entebbe, Uganda

[3]Dept. of Agricultural Production, School of Agricultural Sciences, College of Agricultural & Environmental Sciences, Makerere University, Kampala, Uganda

ABSTRACT

Sorghum domestication probably dates back to the advent of agriculture in sub-Saharan Africa ~10,000 years ago. Today, it is an established understanding that the Sub-Saharan and North East regions of Africa are the primary centers of sorghum diversity. On the African continent, sorghum is grown on 25 million hectares, which constitutes approximately 61% of the world total however African yields average only 0.85 t/ha compared with 4.5 t/ha in North America. The low productivity is associated with; traditional modes of production, low levels of technology adoption, low uptake of improved varieties and

* Corresponding author: Christina.Dixelius@slu.se.

hybrid-seed technology, environmental factors such as unpredicted drought periods and other soil-edaphic constraints, institutional bottlenecks such as research capacity, information barriers, poor infrastructure, marketing hardships, and environmental stresses the major determinants of present sorghum production and yield losses. We have cloned resistance genes (NB-LRR type) to the disease inciting fungi Setosphaeria turcica (St genes) and Colletotrichum sublineolum (Cs genes) causing turcicum leaf blight and sorghum anthracnose, respectively, using a differential transcriptomic approach on Ugandan sorghum genotypes. Presently we are re-sequencing different sorghum genotypes to enhance our understanding of sorghum speciation but more importantly to generate knowledge on additional traits of interest. Wild sorghum species can survive with less water uptake, resist more diseases and insect pests and are presently used to widening the genetic diversity in a pre-breeding initiative. Further, molecular markers are being developed for a genomic selection approach in order to facilitate and speed up the selection processes not least to meet the climate change and emergent food security constrains due to the rapid population growth.

INTRODUCTION

We have chosen to highlight sorghum production in east and central Africa since sorghum stands out as a crop that can withstand drought and thus may become the future staple cereal replacing the temperature sensitive maize in this region of the world. A substantial part of the currently maize growing areas in Africa is predicted to face significant productivity reduction due to climate change and the accelerating temperature increases (World Bank, 2012). Sub-Saharan Africa is also one of the regions of the world with the fastest growing population. It is therefore of utmost importance to speed up breeding processes for crops and livestock, to meet the demand for food and feed, while minimizing the environmental impact of human activities world over. The completion of genome sequences for many important crop species and the rapid development of sequencing and computer technologies provide new strategies to accelerate crop improvements. In this context, sorghum is an outstanding example as a crop with a high quality reference genome and extensive germplasm resources that remain largely unexplored.

Sorghum domestication probably dates back to the advent of agriculture in sub-Saharan Africa (Clark, 1976). Conclusive evidence when sorghum was first brought into some type of cultivation from a prehistoric gathering practice is however lacking. Overall, sorghum seems to have been domesticated in the savannah zone south of the Sahara, certainly earlier than 3,000 B.C. and spread to India and China (Kimber, 2000). Thus, Sub-Saharan and North East regions of Africa are the primary centers of diversity of sorghum with India and China being secondary centers. Australia is considered a tertiary pool of diversity where nineteen wild species indigenous to the continent are found (Lazarides et al., 1991).

Numerous taxonomic systems and evolutionary relationships on grasses have been debated for many years and sorghum is no exception (Dillon et al., 2007). Present taxonomy places the genus *Sorghum*, in *Andropogoneae* (Spangler, 2000; Price et al., 2005). *Andropogoneae* is a diverse tribe, including about 87 genera and 1,060 species, many of which are found in the Katanga province of the current Democratic Republic of the Congo. Based on morphological traits, e.g. grain shape, glumes, and panicle, Harlan and de Wet (1972) developed a simple system to distinguish sorghum species. Today a more robust system for classification of sorghums that takes into account interrelatedness of cultivated and wild sorghums as well as previous taxonomic mistakes and provisions has been developed (Wiersema and Dahlberg, 2007). Under this system five so called sorghum races (*bicolor*, *caudatum*, *guinea*, *kafir*, and *durra*), intermediate crosses between them, and wild related species have been identified and are particularly growing in central and eastern Africa.

Sorghum Production and Constrains in East and Central Africa

In Africa, sorghum is grown on 25 million hectares, which constitutes approximately 61% of the world total, yet yields in Africa average 0.85 t/ha compared with 4.5 t/ha in North America (FAO, 2012). Sorghum is mainly produced as a subsistence crop in most of Sub-Saharan African countries. The farms are small in acreage typically 0.2- 2 ha, have low yields, relatively high production costs and consequently have low returns to investments. To offset their challenges, farmers respond, if possible, by increasing crop acreage. This trend is albeit curtailed by the high population pressure on land and a number of biophysical constraints.

Sorghum is essentially grown in areas where the environment is harsh and growing of other crops is risky. Furthermore, in the eastern and southern African regions, sorghum is grown without any significant use of agricultural inputs such as pesticides, fertilizers and improved seed, unlike other regions of the world. Thus, sorghum productivity remains generally low in most countries (ASARECA, 2008). The low productivity is associated with: traditional modes of production, low levels of technology adoption especially improved varieties with shorter maturity periods, better grain quality, in combination of improved agronomic practices and use of hybrid-seed technology, environmental factors such as unpredicted drought periods and other soil-edaphic constraints, institutional bottlenecks, such as research capacity, information barriers, poor infrastructure, and marketing hardships. Among all constrains however environmental stresses (both abiotic and biotic) are the major determinants of present yield levels and grain qualities.

For cereals, the most critical stages for soil water levels are during early seedling growth and plant establishment and later during grain fill where water deficits can have devastating impact. It has been predicted that arid to semi-arid areas in sub-Saharan Africa are likely to increase between 5-8% by 2080 (Lobell et al., 2008; 2011). This increase in arid areas will result in reduced yields of the major food crops could drop by about 27 % and could bring about an even greater burden to food production. The impact of water stress will be more pronounced with climate change threats, which is expected to increase ambient temperature, rainfall and unpredictable timing of rain-seasons.

Over the decades, small-scale farmers have mined large quantities of nutrients from their fields due to non-replenishment as many of them cannot afford to apply sufficient quantities of manure or fertilizer to the soil. This practice has resulted in a very high average annual depletion rate ~22 kg of nitrogen (N), 2.5 kg of phosphorus (P), and 15 kg of potassium (K) per hectare of cultivated land over the last 30 years equivalent to an annual loss of U.S. $4 billion in fertilizer over 37 African countries, (Sanchez, 2002). Given that most of the African agriculture is rain-fed, production is based on inherent properties of the soil. Sorghum production is consequentially affected by these macronutrient deficiencies.

A growing concern is the expansion of acidic soils with pH $\leq$ 5.2. An estimated 29% of the world's acid soils are distributed across the Sub-Saharan Africa (Eswaran et al., 1997; Vanlauwe and Giller, 2006) where food insecurity is tenuous. In eastern Africa for example, more than 70% of the sorghum cultivation is undertaken on acid soils resulting in yield losses of up to 10%. Most of these soils have high aluminum saturation associated with acidity and/or low available phosphorus (Wortmann et al., 2006; Obura, 2008). A crop growing on Al^{3+} toxic soil is subjected to phosphorus deficiency despite sufficient levels. Low phosphorus availability delays flowering and leads to increased sensitivity to some foliar diseases. Consequently, sorghum growing on acid soils will yield less than their potential because of poorly developed root systems (Anderson and Furlani, 2005).

Striga is a plant genus classified in the family *Orobanchaceae* (former *Scrophulariaceae*) comprising 28 plant species that can be found in Africa, Asia and Australia. A few *Striga* species are plant parasitic with at least two species, *S. hermonthica* and *S. asiatica,* affecting sorghum production. Each *Striga* plant can produce up to 500,000 seeds, which can remain viable in soil for many years. When the seed senses root exudates from a matching host plant, the *Striga* seed germinate and infect its roots (Scholes and Press, 2008). Host plant symptoms such as wilting, stunting and chlorosis are common and similar to a range of soilborne plant pathogens. In some sorghum growing districts of East Africa *Striga* is estimated to be present in up to 80% of the fields, causing yield losses of between 60-85% (Parker, 2009). *Striga* tolerant sorghum varieties have been developed, but many are giving lower yields than the *Striga* susceptible varieties. New promising varieties with *Striga* tolerance have been developed that are more high yielding (Ejeta, 2007). Alternatively, trap cropping involving planting of *Celosia argentea* (the "Striga chaser") between sorghum plants to induce *Striga* seed germination but no further host colonization has shown positive effects. Thus, application of an integrated approach that combines breeding and other control measures is being advocated for (Rich and Ejeta, 2008).

The most important arthropod pests of sorghum include sorghum midge *Contarinia sorghicola*, sorghum shoot fly *Atherigona soccata* which causes substantial losses in late and off-season sorghum along with a number of stem borers including (*Chilo partellus, Buseola fusca, Diatraea sp., Eldana saccharina, Ostrinia nubilalis, Sesamia sp.* and *Buseola sorghuda sp.* (Guo et al., 2011). *Chilo partellus* is found mainly in the semi-arid areas of East Africa while *Sesamia calamitis* and *Buseola sorghida* are distributed throughout sorghum growing areas of Africa (Kfir, 1997). Sorghum shoot fly and stem borers are endemic in most parts of Sub-Saharan Africa (Gitau et al., 2007a; 2007b). Another important pest of the sorghum is the gall midge, which affects flowers, damaging inflorescence. Some varieties are more tolerant than others to this insect pest. More recently, the sorghum chaffer beetle *Pachnoda interrupta* (Coleoptera: *Scarabaeidae*), a non-migratory species, has been observed to affect sorghum production in Ethiopia (Bengtsson et al., 2009).

The major fungal diseases that affect sorghum include: turcicum leaf blight, (*Exserohilum turcicum,* teliomorph: *Setosphaeria turcica*), downy mildew (*Peronosclerospora sorghi*), anthracnose (*Colletrotrichium sublineolum*), grey leaf spot (*Cercospora sorghi*), and a sorghum smut complex including covered kernel smut (*Sporisorium sorghi*), loose smut (*Sphacelotheca cruenta)* and long smuts (*Tolyposporium entrenbargii*) (De Vries and Toeniessen, 2001; Leslie, 2002). These mainly foliar diseases reduce plant photosynthetic area further reducing yield and, at the same time compromise grain quality. For the hybrid-seed initiatives started in the region, the ergot fungus (*Claviceps africana*) posses the biggest threat. Ergot affects unfertilized ovaries and therefore is a great problem to male sterile female genotypes. Ergot alkaloid poisoning is a threat to both cattle and humans. The high incidence of ergot accounts in part to low development of hybrid sorghum technology in many part of Africa.

Birds are one of the most important vertebrate pests of sorghum worldwide. They are capable of inflicting heavy losses causing huge economic losses. The most notorious species is the most abundant wild bird, the red-bill Quelea (*Quelea quelea*). Earlier, sorghums had higher tannin levels, which caused an offensive flavor and was used advantageously to deter birds. These high-tannin sorghums are still grown where birds may cause significant losses.

Other components such as socio-economic factors are in general related to marketing and access to improved seeds and supplies for improving production and products (Ebiyau et al., 2000). These factors include among others: labor shortages, little or no marketing, poor seed systems, which minimize access to improved high yielding varieties. The quality of grains also has implications on meeting international standards for grain exports. There is also a need to diversify the use of grains especially when coupled to small livestock and poultry production as human food. Increasingly in the sub-region there has been a surge in demand for low tannin varieties especially by the bottled bear industry in Uganda, Kenya and Tanzania. There are also regional efforts to promote the use of sorghum grain to manufacture various malted drinks and beverages. Currently the demand for low tannin sorghum, the type need for various industrial processes is high. In Kenya for example the demand for low tannin sorghum is about 32,000 tons with farmers supplying only one third. Similar amounts are requested by large breweries in Nigeria with for example just one mill processing 30,000 tons into malt for beer production. In South Africa sorghum accounts for just below 15% of animal feeds. More recently to reduce pressure on demand for maize bran in feed manufactures, attention is being drawn to the use of sorghum and an increased selection of value added sorghum products.

SORGHUM CULTIVATION IN UGANDA

In Uganda, sorghum is the most important cereal after maize, a crop that is affected by temperatures above 30°C grown under otherwise regular rainfed conditions. Farmers are aware of their environment and drought tolerance among cultivated crops is an important attribute. Sorghum is mostly grown in the drier parts of the country but has some niches even in wet areas especially in the highlands (Airnembabazi et al., 2005). Varieties, mainly of *S. bicolor* subsp. *bicolor*, are adapted to both lowland and highland areas. Lowland sorghum is the most widely grown type especially in the semi-arid area, which covers mostly the eastern and parts of the northern Ugandan regions. This geographic area is characterized by erratic and unpredictable rainfall (500-700 mm per year) and the soils are sandy-loams of medium to low fertility. There are two sorghum production seasons in Uganda and neighboring countries. Sowing is determined by the onset of the rain periods thus the main sowing times are in March and August, respectively, followed by a harvest period in July to August, and a second harvest February to March. Sorghum mono-cropping is the common farming practice in most parts of the country. Farmers sometimes practice intercropping with legumes especially cowpea, and pigeon pea. There is virtually little mechanization except for use of ox-ploughs during land preparation and most other operations are done manually including weeding. Broadcasting is the common traditional method of sowing especially with landraces but with improved varieties, farmers are now practicing row planting.

A number of new varieties including Epuripur, Sekedo, LuluD and Seredo among others have been released since 1995 and more recently SesoI SesoII and SesoIII. All developed at the National Semi Arid Resources Research Institute (NaSARRI) situated at Serere in Soroti, which is a Ugandan institute within the National Agricultural Research Organization (NARO), that has a history of being the center of sorghum breeding for East Africa. Despite the availability of improved varieties, still over 80% of sorghum production is constituted by landraces. Besides socio-economic factors, farmers are quite often fastidious with particular traits like grain taste, color, size and texture of flour. In highlands of south western Uganda and highlands of Burundi and Rwanda, sorghum is produced at high altitudes of 1,890 meter above sea level, the crop takes up to six months, whilst the dryer regions experience receive rains for under 50 days implying short duration varieties are needed, yet most farmers grow long duration cultivars. Thus, clear economic benefits are required for new materials and technologies to become successfully implemented.

Constrains affecting yields are similar to the rest of central and eastern African countries. The incidence and severity of these challenges are compounded by the varied agro-ecological zones where sorghum is produced (Ramathani et al., 2011). In districts of Tororo and Pallisa of eastern Uganda, *Striga* is estimated to infest up to 80% of the fields and causes an estimated yield loss of 60-85%. Sorghum shoot fly and stem borers are endemic in most parts of Uganda (Gitau et al., 2007a) and cause substantial losses particularly in late and off-season sorghum. Larvae from both sorghum midge and shoot fly feed on the seed embryos resulting in not only reduced yields but also large post-harvest losses. The mixture of maize and sorghum varieties, and related crops like pearl millet and a plethora of wild grass species leads to a constant high pest and disease pressure i.e. insect or fungal multiplications on alternate plant host species and eventually to so-called host-jumps of important pathogens. The first sign of such event has been reported on *S. turcica*, where sorghum derived fungal isolates also are capable to infect and cause disease on maize (Ramathani et al., 2011). This type of incidence is important to monitor and adequate measures must be taken in time to avoid further escalating constrains on food security in the region or further spread as in case of stem rust Ug99 on wheat (Wanyera et al., 2006). In theory a mixture of susceptible and resistant plant genotypes often found in genetic diversity centers slows down disease progress but that is rarely seen in this geographic region. One example is elephant grass (Uganda grass or *Pennisetum purpureum*), important as cattle feed, and a grass that attracts stem borers from maize but at the same time works as a reservoir for the insects ready to attack any sorghum plants growing nearby because they have sweeter stems compare to maize due to its closer relatedness to sugar cane. Overall, multiple infections are common making Uganda unique in terms of levels of epidemics. Thus, deploying plants with high levels of multiple resistance traits are desirable and therefore highly prioritized in national breeding programs.

The Sorghum Genome and Future Breeding Opportunities

Sorghum was introduced to America during the 1800s or possibly earlier (Smith and Fredriksen, 2000) and became important in the semiarid plains of the central USA (Menz et al., 2004). Most of the American germplasm derive from either western Africa or the Sudan-Ethiopian region, and much breeding was based on several *zera-zera* sorghums. These genotypes were found useful in that they are less sensitive to photoperiod, have good general combining ability and excellent grain quality, and were at that time resistant to several diseases and insect pests (Rai et al., 1999). Exploitation of F_1 hybrids started in the mid-1950s, with the development of the cytoplasmic male sterility system A1 (Quinby, 1974). BTx623 is a white-seeded, inbred line developed by breeders at Texas A&M University and released in 1976 (Miller, 1977). BTx623 is a widely used maintainer line in the B-cytoplasm system and is classified as a *S. bicolor kafir/zera-zera* derivate. The reference sorghum genome sequencing data is based on the BTx623 genotype.

Plant genome size varies hugely between species and the sorghum genome of 730 Mbp is relatively small compared to other crop plants. For example, its close relative, maize, has a genome of 2,500 Mbp (Schnable et al., 2009). In 2009, Paterson and co-workers sequenced 98% of the sorghum genome revealing a genome content of 27,640 bona fide protein-coding genes based on 34,496 gene models. Putative function has been designated to genes based on homology searches and conserved domain architecture allowing comparisons to be made with other genomes (www.phytozome.net). The sorghum genome contains of 55% retrotransposons, which largely explain the relative net genome expansion compared to rice with a genome size of 389 Mbp (The International Rice Genome Sequencing Project, 2005). Further, the sorghum genome has experienced ancient duplications, today reflected as 57.8% collinear gene models compared to rice. Interestingly, as many as 9,503 orthologous gene families were found between sorghum, rice, Arabidopsis and Poplar genomes. In comparison, 5,337 gene families are shared between sequenced monocots and potato (Xu et al., 2011). Various tools are now under development for grass species such as the integrative database GramineaeTFDB for putative transcription factors (Moshida et al., 2011). A consensus sorghum map with QTLs from a wide range of studies, in order to exploit valuable markers for crop improvements has also been compiled (Mace and Jordan, 2011). Sorghum and sugarcane (*Saccharum officinarum*) are close relatives among cultivated cereals (Figure 1). Their evolutionary divergence is estimated as occurring as recently as ~ 5 million years ago (mya) with maize having separated ~ 15 mya (Paterson et al., 2004; 2010). The evolution of the grass family, was profoundly influenced by a whole genome duplication that occurred ~70 mya, based on comparisons between rice, sorghum, maize and *Brachypodium* genomes (Yu et al., 2005; Wang et al., 2011). Genome data suggest further, that paleo-duplicated gene pairs in sorghum and rice remained extant in about 17% of the cases. Recent duplications of chromosomal segments are particularly found on rice chromosomes 11 and 12, and corresponding regions on chromosome 5 and 8 in sorghum.

In the *S. bicolor* genome, 211 nucleotide binding site – leucine-rich repeat (NB-LRR) encoding potential resistance (R) proteins are present, which is approximately half the number found in rice and slightly more compared to Arabidopsis. Chromosome 5 in the sequenced BTx623 sorghum genotype showed the highest abundance of potential *R*-genes, 62, compared to 30, on chromosome 9. It is postulated that the high numbers of potential *R*-genes in plant genomes and their large sequence diversity are essential evolutionary factors in the surveillance machinery to resist pathogen attacks.

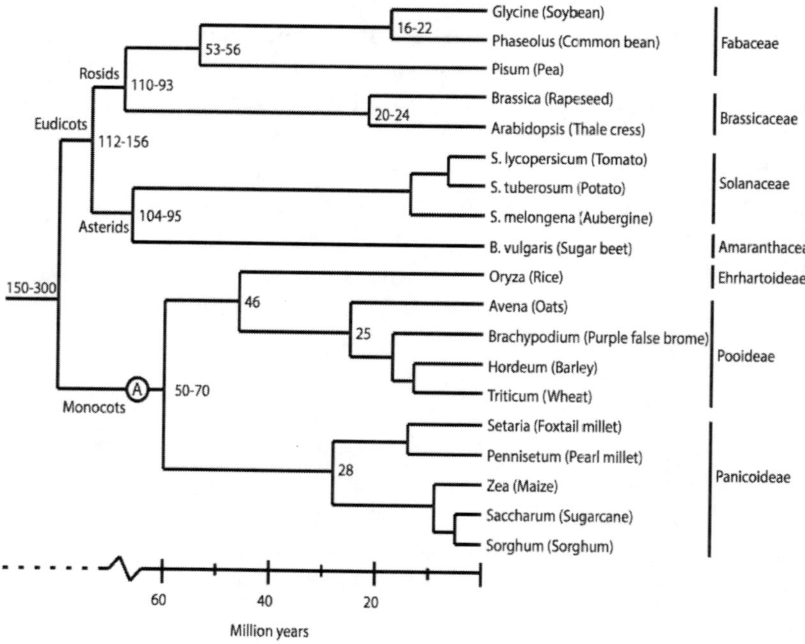

Figure 1. The estimated dates for evolutionary splits in the “green linage” of the upper part of the “tree of life” is as following: gymnosperms (trees) 250 mya, and for 150-300 mya, dicots (eudikots) and monocots (grasses) diverged from each other Dates are based on data by Yang et al. (1999), Wikström et al. (2001), Wang et al. (2009) and Moore et al. (2010). *A* indicates a whole genome duplication event, ~70 mya Salse et al., 2008; Wang et al., 2011).

We have identified *R*-genes (CC-NB-LRRs) to the disease inciting fungus *Setosphaeria turcica* (*St* genes) and *Colletotrichum sublineolum* (*Cs* genes) causing sorghum anthracnose (Martin et al., 2011; Biruma et al., 2012). The six *St* genes in sorghum form three gene pairs in a cluster on chromosome 5. *St* gene orthologs were also found in clusters when searching the rice, maize, foxtail millet and *Brachypodium* genome databases. The data retrieved suggest that the ancestor of rice and sorghum likely had a copy of four out of the six genes, whereas one *St* gene pair is a result of a more recent genome duplication event. In a parallel study, two *Cs* loci were found on chromosome 9. A phylogenetic comparison showed that Cs2 proteins cluster closest to the St proteins, harbor a WD40 domain at their C-terminal ends, whereas the *Cs1* gene pair is more distantly related (Figure 2).

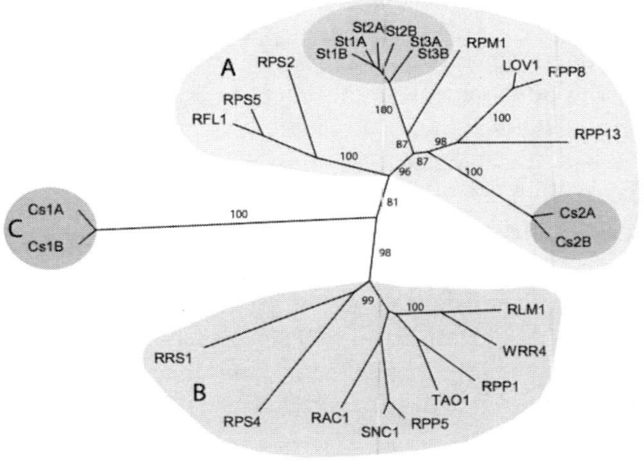

Figure 2. Unrooted maximum-likelihood phylogram inferred from nucleotide binding (NB) and leucine rich repeat (LRR) domains, of four resistance proteins to *C. sublineolum* (Cs) and six resistance proteins to *S. turcica* (St) in sorghum, compared with Arabidopsis NB-LRR resistance proteins with known function. LR-ELW values above 75% are shown. Labeling is as follows: (A) NB-LRR resistance proteins with NB-LRR proteins with a coiled-coil (CC) domain at the N- terminal end. (B) NB-LRR resistance proteins with a Toll/Interleukin-1 receptor (TIR) at the N- terminal end. (C) Two of the sorghum proteins did not clustering with any Arabidopsis protein candidate. The phylogenetic analysis was performed as outlined in Martin et al. (2011).

A central event of angiosperm evolution is the extent of genome duplication (Smith et al., 2002). Whole genome duplications, revealed in all plant genomes sequenced so far, seems to play major roles in shaping the genomes and contributing to novel characteristics like nitrogen fixation in legumes (Young et al., 2011). Gene order conservation in vertebrates is still evident after several hundreds of millions of years of divergence. Still rather high numbers of lineage-specific genes can be found in each eukaryotic genome. For example, 3,983 gene families in sorghum are absent from the dicot genomes known to date and 1,053 are absent from the rice genome. This observation suggests that we have at least one thousand sorghum specific gene families to explore together with selected basic plant or grass genes.

Crop genome information generates a suit of new experimental and computational approaches to accelerate crop improvements. Sorghum is a remarkable versatile crop that can be productive under adverse conditions. One interesting aspect is to expand uses of perennial grains to further decrease nutrient leakage (Glover et al., 2010). A close relative, Johnson grass, (*S. halepense*), one of the world´s most noxious weeds, a perennial that spread by seeds and rhizomes (Holm et al., 1977). Both perennial and rhizomatousness traits are interesting to understand and exploit to various crops not least forage and biomass crops or genotypes. In this context sorghum genotypes with high sugar content in stems (sweet sorghums) have received new attention. Sorghum is already the number two crop used in ethanol production after maize in USA and has received recent attention in Sub-Saharan Africa. Ethanol is the single largest market for US grain sorghum producers (www.sorghumgrowers.com). The impact of sweet sorghum is expected to increase significantly since few advanced varieties have reached the market. Genome sequencing has identified ~1,500 genes that differ between sweet and grain sorghum and this information is now implemented in breeding improvements (Zheng et al., 2011).

Ongoing Work and Future Perspectives

We have highlighted a number of constraints in East and Central Africa when it comes to sorghum production but there is also a wealth of possibilities for new innovations and sorghum products that could strengthening the bio-economy, and thereby increase the incomes in the entire society. To make that change to happen, much more advanced varieties need to be available for the farmers. In parallel, technical adjustments like those that has made possible to replace malted barley with sorghum need to be developed and investments in many other new enterprises.

Together with collaborators we are presently re-sequencing a number of *S. bicolor* varieties of importance for East Africa together with related wild sorghum species (*S. verticilliflorum, S. arundinaceum,* and *S. propinquum*). *S. arundinaceum* and *S. verticilliflorum* that harbor traits such as high resistance to sorghum midge, and shoot fly as well as tolerance to several fungal diseases and tolerance to drought. Beside problems with insects and fungal diseases, sorghum has much lower protein digestibility compared to other cereals, even if the sorghum grains are cooked (Belton et al., 2006). Thus, to improve the nutritional value of sorghum its seed storage composition needs to be altered. The storage protein in sorghum endosperm is called kafirin and is divided into four subclasses, α-, β-, γ- and δ-kafirin. The β- and γ- kafirins are the ones particularly contributing to the reduced digestibility. Interestingly, a β-kafirin null mutant together with new alleles for the three other kafirin gene classes was recently found in the wild sorghum *S. propinquum* (Laidlaw et al., 2010). It should however be pointed out that sorghum do not have gluten in the endosperm, thus producing gluten free foods could be a potential new niche not least for different export products. To further advance the use of the expanding single nucleotide polymorphism (SNPs) and genome sequence knowledge, a new initiative has been taken to develop a platform for genomic selection. Genomic prediction combines marker data covering the whole genome with phenotypic and pedigree data (when available) in an attempt to increase the accuracy of the prediction of breeding and genotypic values (Morell et al., 2012; Morris et al., 2013). A strategy expected to revolutionize breeding in the next decade. There is a strong regional wish to expand the platform and include a number of genotypes harboring important traits for increased sorghum production, for example resistance to *Striga*, aluminum tolerance and more complex yield-related traits. The future prospects for sorghum as a natural resource with multiple use and new products is exceedingly promising but breeding for East and Central African demands require skilled people in bioinformatics and new genetic advancements to work hand-in-hand with plant breeders in dialogue with end-users to quickly and cost-effectively generate materials for the rapid growing population.

ACKNOWLEDGMENTS

This work was supported by the Swedish Ministry of Foreign Affairs as part of its special allocation to SLU on global food security, the Sida grant

SWE-2005-453, the BIO-EARN & BIO-INNOVATE programs, and the Swedish University of Agricultural Sciences (SLU).

REFERENCES

Ainembabazi, J.H., Bashaasha, B., Mugisha, J., Pende, J. and Hyuha, T.S. (2005) Technological change in sorghum production in Eastern Uganda. African Crop Science Conference Proceedings, 7, 947-954.

Anderson, R.M. and Furlani, P.R. (2005) Tolerance to aluminum toxicity by tropical leguminous plants used as cover crops. *Braz. Arch. Biol. Techn.* 48, 309-317.

ASARECA (2008) Strategic Choices and Research Priorities for the ASARECA Sub-Region: Food Crops, Livestock, Natural Resources Management, Policy and Information. Executive Summaries, edited by Seyfu Ketema, 2008, ASARECA, Entebbe, Uganda.

Belton, P.S., Delgadillo, I,. Halford, N.H. and Shewry, P.R. (2006) Kafirin structure and functionality. *J. Cereal Sci.* 44, 272-286.

Bengtsson, J.M., Walde-Hawariat, Y., Khbaish, N., Negash, M., Jembere, B., Seyoum, E., Hansson, B.S., Larsson, M.C. and Hilbur, Y. (2009) Field attractants for *Pachnoda interrupta* selected by means of GC-EAD and single sensilium screening *J. Chem Ecol.* 35, 1063-1076.

Biruma, M., Martin, T., Fridborg, I., Okori, P. and Dixelius, C. (2012) Two loci in sorghum with NB-LRR encoding genes confer resistance to *Colletotrichum sublineolum*. *Theor. Appl. Genet.* 124,1005-1015.

Clark, J.D. (1976) Prehistoric populations and pressures favouring plant domestication in Africa. In: JR Harlan, JMJ de Wet and ABL Stemler eds. Origins of African Plant domestication. Mouton. The Hague, The Netherlands, pp 67-195.

De Vries, J. and Toenniessen, G. (2001) Securing the harvest. Biotechnology and breeding methods for African Crops. The Cromwell Press, Wiltshire, UK. 208 pp.

Dillon, SL., Shapter, FM., Henry, RJ., Cordeiro G., Izquirdo, L. and Lee, L.S. (2007) Domestication to crop improvement: genetic resources for Sorghum and Sacharum (*Andropogoneae*). *Ann. Bot.* 100, 975-989.

Ebiyau, J., Oryokoti, J., Esele, J.P., Amongin, B. and Okwi, B. (2000) Performance of advanced sorghum lines under diverse environments'. *Ugandan J. Agric. Sci.* 5, 15-19.

Ejeta, G. (2007) Breeding for *Striga* resistance in sorghum: exploitation of an intricate host-parasite biology. *Crop Sci.* 47, S216–S227.

Eswaran, H., Almaraz, R., Reich, P.F. and Zdrull, P.F. (1997) Soil quality and soil productivity in Africa. *J. Sustainable Agric.*10, 75–94.

Food and Agriculture Organization of the United Nations (FAO) (2012) www.faostat.fao.org

Gitau, C.W., Gundersen-Rindal, D., Pedroni, M., Mbugi, P.J. and Dupas, S. (2007b) Differential expression of the CrV1 haemocyte inactivation-associated polydnavirus gene in the African maize stem borer *Busseola fusca* (Fuller) parasitized by two biotypes of the endoparasitoid *Cotesia sesamiae* (Cameron). *J. Insect Physiol.* 53, 676-684.

Gitau, C.W., Ngi-Song, A.J., Otieno, S.A. and Overholt, W.A. (2007a) Host preference of *Xanthopimpla stemmator* (Hymenoptera: Ichneumonidae) and its reproductive performance on selected African Lepidopteran stem borers. *Biocontrol Sci. Techn.* 17, 499-511.

Glover, J.D., Reganold, J.P., Bell, L.W., Borevitz J., Brummer, E.C. *et al.* (2010) Increased food and ecosystems security via perennial grains. *Science*, 328, 1638-1639.

Guo, C., Cui, W., Feng, W., Zhao, J. and Lu, G. (2011) Sorghum insect problems and managements. *J. Integrative Plant Biol.* 53, 178-192.

Harlan, J.R. and de Wet, J.M.J. (1972) Simplified classification of cultivated sorghum. *Crop Sci.* 12, 172-176.

Holm, L.G., Plucknett, D.L,, Pancho, J.V. and Herberger, J.P. (1977) The worlds worst weeds: distribution and biology Univ. Press of Hawaii, Honolulu HI, US.

Kfir, R. (1997) Natural control of the cereal stemborers *Busseola fusca* and *Chilo partellus* in South Africa. *Insect Sci. Appl. 17*, 61-67.

Kimber, C.T. (2000) Origins of domesticated sorghum and its early diffusion into India and China. In: Sorghum: origin, history, technology and Production. C.W Smith & R.A. Frederiksen (eds.), John Wiley & Sons, New York, 3-98.

Laidlaw, H.K.C., Mace, E.S., Williams, S.B., Sakrewski, K., Mudge, A.M., Prentis, P.J., Jordan, D.R. and Goodwin, I.D. (2010) Allelic variation of the β-, γ- and δ-kafirin genes in diverse *Sorghum* genotypes. *Theor. Appl. Genet.* 121,1227-1237.

Lazarides, M., Hacker, J.B. and Andrew, M.H. (1991) Taxonomy cytology and ecology of indigenous Australian sorghums, Sorghum Moench, Adropogoneae, Poaceae. *Austr. Syst. Bot.* 4, 591 – 636.

Leslie, J.F. (2002) Sorghum and millets diseases. Wiley-Blackwell.

Lobell, D.B., Burke, M.B., Tebaldi, C., Mastrandrea, M.D., Falcon, W.P. and Naylor, R.L. (2008) Prioritizing climate change adaptation need for food security in 2030. *Science*, 319, 607-610.

Lobell, D.B., Schlenker, W. and Costa-Roberts, J. (2011) Climate trends and global crop production since 1980. *Science*, 333, 616-620.

Mace, R.S. and Jordan, D.R. (2011) Integrating sorghum whole genome sequence information with a compendium of sorghum QTL studies reveal uneven distribution of QTL and of gene-rich regions with significant implications for crop improvement. *Theor. Appl. Genet.* 123, 169-191.

Martin, T., Biruma, M., Fridborg, I., Okori, P. and Dixelius, C. (2011) A highly conserved NB-LRR encoding gene cluster effective against *Setosphaeria turcica* in sorghum. *BMC Plant Biol.* 11, 151.

Menz, M.A., Klein, R.R., Unruh, N,C., Rooney, W.L., Klein, P.E. and Mullet, J.E. (2004) Genetic diversity of public inbreds of sorghum determined by mapped AFLP and SSR markers. *Crop Sci.* 44, 1236–1244.

Miller, F.R. (1977) Release of A & BTx622, 623, 624. Report of technical committee on seed release and increase. Tex. Agric. Exp. Stn., College Station TX. USA.

Moore, M., Soltis, P., Bell, C., Burleigh, G. and Soltis, D. (2010) Phylogenetic analysis of 83 plastid genes further resolves the early diversification of eudicots. *Proc. Natl. Acad. Sci. USA*, 107, 4623–4628.

Morell, P.L., Buckler, E.S. and Ross-Ibarra, J. (2012) Crop genomics: advances and applications. *Nat. Rev. Genet.* 13, 85-96.

Morris, G.P., Ramu, P., Deshpande, S.P., Hasth C.T., Shah, T. *et al.* (2013) Population genomic and genome-wide association studies of agroclimatic traits in sorghum. *Proc. Natl. Acad. Sci. USA*, 110:4453-458.

Moshida, K., Yoshida, T., Sakurai, T., Yamagushi-Shinozaki, K., Shinozzaki, K. and Tran, L-S. P. (2011) *In silico* analysis of transcription factor repertoires and prediction of stress-responsive transcription factors from six major *Gramineae* plants. *DNA Res.* 18, 321- 332.

Obura, P. (2008) Effect of soil properties on bioavailability of aluminum and phosphorus in selected Kenyan and Brazilian acid soils. PhD thesis Purdue Univ. USA

Parker, C. (2009) Observations on the current status of *Orobanche* and *Striga* problems worldwide. *Pest Manag. Sci.* 65, 453-459.

Paterson, A.H., Bowers, J.E. and Chapman B.A. (2004) ancient polyploidization predating divergence of the cereals, and its consequences for comparative genomics. *Proc. Natl. Acad. Sci. USA*. 101, 9903-9908.

Paterson, A.H., Bowers, JE., Bruggmann, R., Dubchak, I., Grimwood, J. *et al.* (2009) The *Sorghum bicolor* genome and the diversification of grasses. *Nature,* 457, 551-556.

Paterson, A.H., Freeling, M., Tang, H. and Wang, X. (2010) Insights from the comparison of plant genome sequences. *Annu. Rev. Plant Biol.* 61, 349-372.

Price, H.J., Dillon, SL., Hodnett, G., Rooney, W.L., Ross, L. and Johnston, J.S. (2005) Genome evolution in the genus *Sorghum* (Poaceae). *Ann. Bot.* 95, 219-227.

Quinby, J.R. (1974) Sorghum improvement and the genetics of growth. Texas A&M Univ. Press. College Station, TX, USA.

Rai, K.N., Murty, D.S., Andrews, D.J. and Bramel-Cox, O.J. (1999) Genetic enhancement of pearl millet and sorghum for the semi-arid tropics of Asia and Africa. *Genome*, 2, 617-628.

Ramathani, I., Biruma, M., Martin, T., Dixelius, C. and Okori, P. (2011) Disease severity, incidence and races of *Setosphaeria turcica* on sorghum in Uganda. *Eur. J. Plant Pathol.* 131, 383–392.

Rich, P.F. and Ejeta, G. (2008) Towards effective resistance to *Striga* in African maize. *Plant Sign. & Behav.* 3, 618-621.

Salse, J., Bolot, S., Throude, M., Jouffe., Piegu, B., Quraishi, U.M., Calcagno, T., Cooke, R., Delseny, M. and Feuillet, C. (2008) Identification and characterization of shared duplications between rice and wheat provide new insight into grass genome evolution. *Plant Cell*, 20, 11- 24.

Sanchez, P. A. (2002) Soil fertility and hunger in Africa. *Science*, 295, 2019-2020.

Schnable, P., Ware, D., Fulton, R., Stein, J., Wei, F. *et al.* (2009) The B73 maize genome: complexity, diversity and dynamics. *Science,* 326, 1112-1115.

Scholes, J.D. and Press, M.C. (2008) *Striga* infestation of cereal crops – an unsolved problem in resource limited agriculture. *Curr. Opin. Plant Biol.* 11, 180-186.

Smith, C.W. and Frederiksen, R.A. (2000) Sorghum: Origin, history, technology and production. John Wiley and Sons.

Smith, S.F, Snell, P., Gruetzner, F., Bench, A.J., Haaf, T. *et al.* (2002) Analyses of the extent of shared synteny and conserved gene orders between the genome of *Fugu rubripes* and human 20q. *Genome Res.* 12, 776-784.

Spangler, R.E. (2000) *Andropogoneae* systematics and generic limits in Sorghum. In S.W.L. Jacobs and J. Everett (eds.). Grasses: Systematics and

Evolution. International Symposium on Grass Systematics and Evolution. CSIRO Publishing, Collingwood, Victoria, Australia. pp. 167-170.

The International Rice Genome Sequencing Project (2005) The map-based sequence of the rice genome. *Nature,* 436, 793-800.

Vanlauwe, B. and Giller, K.E. (2006) Popular myths around soil fertility management in sub-Saharan Africa. *Agric. Ecosyst. Environ.* 116, 34-46.

Wang, H., Moore, M.J., Soltis, PS., Bell, C.D., Brockington, S.F., Alexandre, R., Davis, C.C., Latvis, M., Manchester, S.R. and Soltis, D.E. (2009) Rosid radiation and the rapid rise of angiosperm-dominated forests. *Proc. Natl. Acad. Sci.* 106, 3853 -3858.

Wang, X., Tang, H. and Paterson A.H. (2011) Seventy million years of concerted evolution of a homologous chromosome pair, in parallel, in major *Poaceae* linages. *Plant Cell*, 23, 27-37.

Wanyera, R., Kinyua, M.G., Jin, Y. and Sing, R.P (2006) The spread of stem rust caused by *Puccinia graminis f.sp.tritici* with virulence on *Sr31* in wheat in Eastern Africa. *Plant Dis.* 90,113.

Wiersema, J.H. and Dahlberg, J. (2007) The nomenclature of *Sorghum bicolor* (L.) Moench (*Gramineae*). *Taxon,* 56, 941–946.

Wikström, N., Savolainen, V. and Chase, M.W. (2001) Evolution of the angiosperms: calibrating the family tree. *Proc. Biol. Sci.* 268, 2211-2220.

World Bank (2012) Turn down the heat. Why a 4°C warmer world must be avoided. Report, November 2012.

Wortmann, C.S., Mamo, M., Abebe, G., Mburu, C., Kayuki, K.C., Letayo, E. and Xerinda, S. (2006) Sorghum Atlas: Production in five countries of Eastern African. University of Nebraska, Lincoln. 51pp.

Xu, P., Pan, S., Cheng, S., Zhang, B., Mu. D. *et al.* (2011) Genome sequence and analysis of the tuber crop potato. *Nature*, 475, 189-195.

Yang, Y.W., Lai, KN., Tai, P.Y. and Li, W.H. (1999) Rates of nucleotide substitution in angiosperm mitochondrial DNA sequences and dates of divergence between Brassica and other angiosperm lineages. *J. Mol. Evol.* 48, 597-604.

Young ND, Debellé F, Oldroyd GED, Geurts R., Cannon, S. *et al.* (2011) The *Medicago* genome provides insight into the evolution of rhizobial symbioses, *Nature*, 480, 520-524.

Yu, J., Wang, J., Lin, W., Li, S. *et al.* (2005) The genomes of *Oryza sativa*: A history of duplications. *PLoS Biol.* 3, e38.

Zheng, E-Y., Guo, X-S., He, B., Sun, L-J., Peng, Y. *et al.* (2011) Genome-wide patterns of genetic variation in sweet and grain sorghum (*Sorghum bicolor*). *Genome Biol.* 12, R114.

In: Sorghum
Editor: Patricia C. Parra
ISBN: 978-1-62618-351-3

Chapter 3

THE HEALTH BENEFITS OF SORGHUM GRAIN FOR POULTRY AND PIGS

***M. Mabelebele*[1,2] *and P. A. Iji*[2]**
[1]School of Agricultural and Environmental Sciences,
University of Limpopo, South Africa
[2]School of Environmental and Rural Science,
University of New England, Australia

ABSTRACT

Sorghum is the fifth most important grain crop after wheat, rice, maize and barley. Sorghum is cultivated for food and beverages for humans, and for feed and fodder for animals in America, Asia, Australia and Africa. Its crude protein content is higher than that of maize whereas its energy value is rated as high as that of maize. Sorghum tolerates poorer growing conditions than maize, so is more abundant in marginal regions, where it is used to feed humans and animals. Sorghum may confer some health benefits on consumers due to the nature of its carbohydrates but also the presence of other chemicals, which ordinarily are regarded as antinutrients in animal nutrition, e.g. polyphenols or tannins and polycosanols. When fed to humans, these factors may be beneficial. Tannins would reduce feed intake and digestive enzyme activities, to minimize weight gain while polycosanols, present in the waxy layer, are known to reduce the concentrations of low-density lipoproteins and improve cardiac function. Sorghum is also free from gluten and therefore useful for people suffering from coeliac disease. Some of these benefits are also noticeable in animals although could be negative to productivity. Sorghum is therefore not seen to be an ingredient for controlling animal health and its use as such has not received significant attention in animal nutrition. This chapter discusses the direct health benefits associated with feeding sorghum grains to poultry and pigs. It will also link changes in animal product quality to health benefits of the human consumer.

Keywords: Poultry, pigs, sorghum, health benefits, nutritional properties

1. INTRODUCTION

Sorghum is an important food and feed ingredient in many parts of the world. Certain agronomic characteristics, such as drought tolerance, allow sorghum to be grown in diverse geographic areas (Sedghi *et al.*, 2011). It out-performs most cereals due to its ability to endure diverse environmental stressors and thus, can be economically produced (Awika, 2004). Over the years, different sorghum cultivars have been developed and are known to have different physical and chemical characteristics and large variations in digestibility (Neucere and Sumrell, 1980). When used for animal feeding, sorghum has a feeding value of 95 % of that of maize even though it is priced less than maize in many markets. Sorghum is used as a staple food for humans in many parts of the world. Its consumption as refined and whole grains has many benefits to human.

Sorghum contains various phytochemicals (Figure 1). These are secondary plant metabolites or integral cellular compounds. Phenols help in the natural defence of plant pests and diseases, while policosanols are mostly compounds of wax and plant oils. Sorghums vary widely in their phenolic composition and content, with both genetics and environment affecting the kind and level of phenolic compounds present (Awika, 2004). Sorghum can be classified as *type I*, which do not have pigmented testa layer or condensed tannins and low levels of phenols, *type II* have a pigmented testa layer and generally have intermediate condensed tannins levels whereas *type III* sorghum also have a pigmented testa and high levels of tannins. However, this classification does not account for the varying levels of major phenolic constituents, especially anthocyanins. The presence of phytochemicals in sorghum has been demonstrated to have health-promoting effects in both humans and animals (Walker, 1999; Burdette, 2003; Awika *et al.*, 2004; Dykes *et al.*, 2006; Mathanghi, 2012). However, the potential of sorghum's phytochemicals and anti-oxidant activities for animal health is not well documented. Therefore, the purpose of this chapter is to review direct health benefits associated with feeding sorghum grains to poultry and pigs. It will also link changes in animal product quality to health benefits of the human consumer.

2. Nutritional Properties of Sorghum Grains

The nutrient composition of sorghum has been well studied and documented (Oyenuga, 1968; NRC, 1984; Etuk, 2008) and indicates that it is a good source of energy, protein, vitamins, and minerals, including trace elements. Whole grain sorghum is known to contain approximately (%) 89-90 dry matter (DM), 8.9-15 crude protein (CP), 3.4 ether extract (EE), 1.5-1.7 ash, 2.1-2.3 crude fibre and 71.7-72.3 nitrogen-free extract on as-fed basis. Sorghum grain has a true metabolisable energy corrected for nitrogen (TMEn) content of between 3003 and 3899 kcal/kg (Sedghi*et al.*, 2011). Sorghum contains low levels of lysine but high tryptophan content relative to maize (Purseglove, 1972; Olomu, 1995). Generally, sorghum is known to contain high levels of fibre and tannins (Tacon, 1995; Aletor, 1999). However, the discussion on nutritional properties is not complete without the non-nutritional properties present in sorghum. Figure 1 illustrates the classification of many types of phytochemicals, which are defined as bioactive non-nutrient plant compounds in fruits, vegetables and grains (Burdette, 2003). The phytochemicals of interest for this chapter are tannins, phenolic acids and flavonoids.

2.1. Tannins

Tannins are usually subdivided into two groups, hydrolysable tannins and proanthocyanidins, which are commonly known as condensed tannins. Condensed tannins, which are more widely distributed than hydrolysable tannins, are oligomers or polymers of flavonoid units. Tannins in sorghum have the ability to bind protein and form insoluble or soluble tannin-protein complexes and also complex with starch, cellulose and minerals. They can also have a large influence on the nutritive value of sorghum fed to both humans and animals. Their role has been identified in several conventional and alternative pulses and legumes that are usually fed to poultry (Kahlon and Shao, 2004). In some parts of Africa, traditional sorghum varieties of moderate tannin content are widely grown and used as staple food and alcoholic beverages. In Southern Africa tannin-rich sorghum is preferred because it sustains farmers for longer while they work in the fields (Awika, 2004). The tannin content of sorghum is often thought to be closely related to darkness of seed colour; however, it is not an adequate indicator of tannin content (Boren and Waniska, 1992). Therefore, quantitave/qualitative methods of determining tannin content are normally employed. It is important to note that majority of the sorghum that is grown in USA, Australia and other parts of the world are tannin-free.

2.2. Phenolic Acids

Phenolic acids are divided into hydroxybenzoic and hydroxycinnamic acids. Phenolic acids that occur in sorghum grains are in free and bound states. Free phenolic acids are located in the pericarp and bound phenolic acids are esterified to cell walls in the endosperm; acid or base hydrolysis is required to release these bound compounds from the cell matrix (Kim *et al.*, 2006). The major phenolic acids in sorghum are ferulic and *p*-coumaric acids (Mattlila *et al.*, 2005). These phenolic acids are known for their strong antioxidant capabilities (Awika *et al.*, 2005). In sorghum, bound phenolic acids account for 85 % of total phenolic acids (Waniska *et al.*, 1989). Phenolic acid levels differ between cereals and the brans tend to contain three times the levels of these compounds in the parent grain (Table 1).

2.3. Flavonoids

These are compounds with C_6-C_3-C_6 skeleton that consist of two aromatic rings joined by a three carbon link. Flavonoids are subdivided by their differences in the generic structure of the heterocyclic C ring as flavonols, flavones, flavanols (catechins), flavanones, anthocyanidins and isoflavonoids (Figure 2). More than 5000 flavonoids have been identified in nature (Yao *et al.*, 2004). These are located in the pericarp of sorghum and all other cereals, with orghum having the widest variety of flavonoids. Anthocyaninsare major flavonoids present in cereals. Sorghum contains unique anthocyanins called 3-deoxyanthocyanins, which lack a hydroxyl group in the third position of the C-ring. This feature increases their stability at high pH compared to common anthocyanin (Gous, 1989; Awika *et al.*, 2004). There is evidence that anthocyanins in sorghum are responsible for the grain's high anti-oxidant activity (Kim *et al.*, 2006).

3. SORGHUM GRAINS IN HUMAN FOOD AND ANIMAL FEED

3.1. Sorghum Grains as Human Food

Sorghum is used in a variety of food in many parts of the world. It is consumed as whole grain or further processed into flour, from which traditional meals are prepared. Other varieties of sorghum are used for food in various parts of the world, for instance, in eastern and southern Africa sorghum with moderate tannin content is consumed as staple food (Awika *et al.*, 2004). There was a recent shift in the use of whole grain sorghum to refined grains in human diet (Slavin, 2004). Earlier grinding mills did not separate the bran and germ from the endosperm, however, such separation is achieved with the use of roller mill. Refined grains have much more starch but fewer phytochemicals than whole grains (Hu, 2003). Inclusion and consumption of whole grains in diets of human has been associated with decreased incidence of cancer, cardiovascular disease, diabetes and obesity (Slavin, 2004). Whole grain sorghum contains many antioxidant compounds, fat- and water-soluble- vitamins and phenolic acids, which can greatly benefit human health.

3.2. Sorghum Grain as an Animal Feed

Approximately 50 % of the sorghum grain produced is fed to livestock, and the response is often comparable to that on maize, for which it is a close substitute. Sorghum has similar feed characteristics to maize and provides as much metabolisable energy, has higher crude protein content but is less digestible. Total digestible nutrients in sorghum are 95 % of those in maize. The nutritional deficiency relative to sorghum is assumed to be due to lower starch availability. The nutritional value of sorghum grain is different for ruminant and non-ruminant animals. In ruminants, sorghum is viewed as a source of carbohydrates (starch), whereas in non-ruminants it is mainly an energy source. When sorghum-based diet is offered to broiler chickens, proteins from the cereal contribute only one third of the total dietary crude protein required (Dowling, 2002).

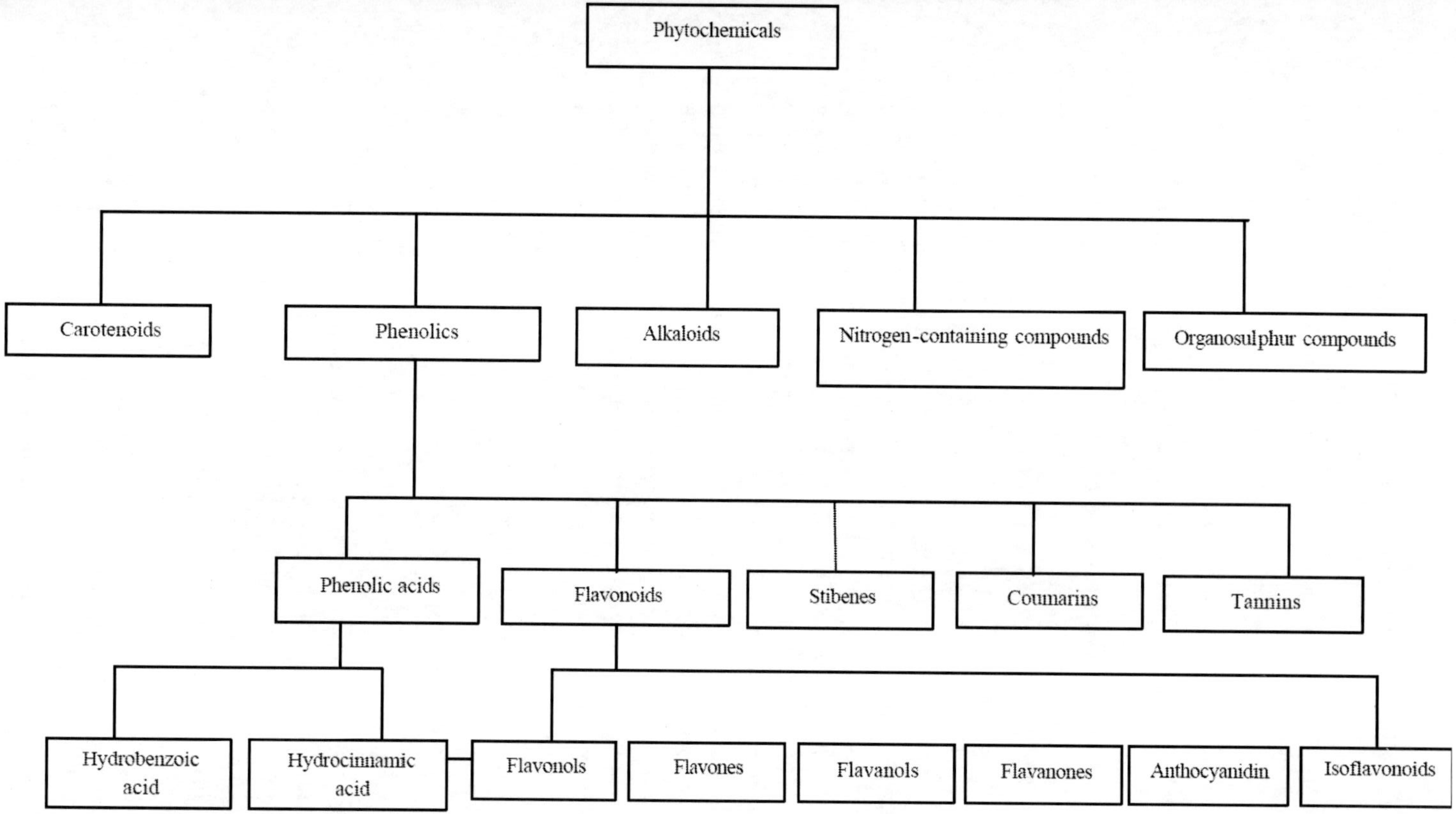

Figure 1. Range of phytochemicals present in fruits, vegetables and grains (Adapted from Liu, 2004).

Depressed growth and feed conversion ratio have been observed when regular sorghum grain high in tannin is fed to poultry diets, however, modern sorghum varieties with lower or no tannin show support more productivity when fed to non-ruminant animals. Sorghum grain has not been shown to have any negative effect on the performance of pigs.

Table 1. Phenolic acid contents of cereal grains

Cereal grain (Whole)	Amount (mg/kg)	References
Barley	450-1346	1, 2
Maize	601	2
Oat	472	2
Pearl millet	1478	3
Rice	197-376	2
Sorghum	**385-746**	4
Wheat	1342	2

References: 1. Holtekj¢len *et al.* (2006). 2. Mattlila et al. (2006). 3 McDonough et al. (2000). 4. Hahn et al. (1983).

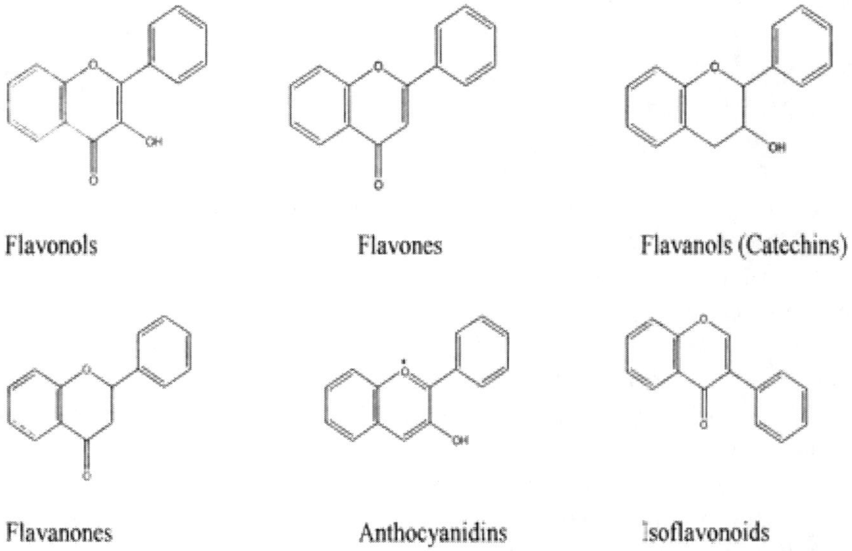

Figure 2. Structure of flavonoids (Adapted from Liu, 2004).

4. The Health Benefits of Sorghum Grains for Broiler and Layer Chickens

4.1. Role and Properties of Phenolic Compounds

It has already been stated above that sorghum is a rich source of phytochemicals, which have the potential to significantly impact animal health. This section discusses in detail the effects of phenols on nutrition and health benefits of broiler and layer chickens. Condensed tannins e.g. proanthocyanidins, may bind to proteins, carbohydrates and minerals thereby affecting the nutrition and functional value of the bound components. The main nutritional concerns about the condensed tannins are their ability to bind strongly to large proteins and thereby reduce their digestibility (Butler, 1992; Dicko *et al.*, 2005). Condensed tanninsmay be anti-nutritional through direct intrusion within the animal body of the digestive processes, or inhibition of hydrolytic enzymes (Nguz *et al.*, 1998).Tannins in sorghum do not cause toxicity problem in animals consuming the grain, but the feed efficiency is compromised. Jacob *et al.* (1996) did not observe any adverse effects in broiler chicks fed diets containing 1.3 % catechins. Ambula *et al.* (2003) also reported that tannin-rich sorghum; containing 5 % catechin equivalents (CE) did not cause adverse effects on the performance of laying hens. However, rabbits fed on diets with high-tannin sorghums (3.5 % catechin equivalents) had significantly reduced weight gain, feed conversion ratio, and slightly higher food consumption when compared to the control. Sorghum containing tannins is not toxic to animals or humans and can be used effectively in feed and foods. The classification of condensed tannins as anti-nutritional factors is now questioned because it is thought to have health benefits for humans (Hagerman *et al.*, 1998; Waniska, 2000, Parr and Bolwell, 2000; Clifford, 2000; Awika, 2004a).

The total phenol content of sorghums is highly correlated with its antioxidant activity (Awika *et al.*, 2003b; Dykes *et al.*, 2005). Such high correlation (r = 0.79 – 0.88) have been reported for red pericarp sorghum (Dicko *et al.*, 2005; Dykes *et al.*, 2005). Sorghum with a pigmented testa increased antioxidant activity due to the presence of tannins, which are more influential antioxidants than monomeric phenolic compounds. There is higher antioxidant activity in sorghum with condensed tannins than in fruits (Table 2). Several health benefits have been reported in humans and may be of similar benefits to poultry.

4.2. The Health Benefits of Sorghum Grains for Broiler and Layer Chickens

The consumption of whole grain sorghum has been shown to significantly decrease mortality from cardiovascular diseases (Slavin *et al.*, 2000; Anderson, 2003). This may be due to the phytosterols present in sorghum bran, together with other components of the grains, including polyphenols and fibre. Klopfenstein *et al.* (1981) reported the cholesterol-lowering effect of low-tannin sorghum grain fed to guinea pigs at 58 % of diet while Cho *et al.* (2000) observed an increased level of high-density lipoprotein (HDL) cholesterol without change in total cholesterol when whole sorghum was fed to rats at 30 % of diet. Broiler chicks fed low-tannin sorghums (0.6–0.9 % CE) had similar weight gain and feed efficiency compared to those on corn (Ambula *et al.*, 2001). However, these authors observed an average of 50 % reduction in weight gain when high-tannin sorghums (2.7–3.5 % CE) were fed to broiler chicks. Such data suggest that there is a threshold for the inhibitory effects of sorghum tannins, and that animal species are affected differently.

Jambubathan and Mertz (1973); Featherson and Rogler (1975); Cousins *et al.* (1981); Lizardo*et al.* (1995); Al-Mamary*et al.* (2001), and Muriu *et al* (2002) reported on reduced weight gain of animals such as rats, pigs, rabbits and poultry fed high-tannin sorghum. Tannin-rich sorghum reduces the nutritive value of diets by binding to food proteins (Haslam, 1974; Hagerman and Butler, 1981) and carbohydrates (Naczk and Shahidi, 1997) so that they cannot be broken down by digestive enzymes. The effects of sorghum tannins on animal weight gain depend on levels fed as well as animal species. The property of slow digestibility and nutrient release from the tannin-complexed food matrix should be investigated, in order for sorghum to be used in human diet to help fight obesity. This is an effect that may be useful in controlling carcass fat, especially for meat animals that are kept for longer than the conventional period of rearing. Broiler chickens, for example, may be kept for such long periods to increase the yield of select meat parts such as breast and drumsticks. Gomez-Cordovez*et al.* (2001) showed that sorghum tannins had anti-carcinogenic activity against human melanoma cells, as well as positive melanogenic activity (melanogenesis is believed to help protect human skin against UV irradiation damage). According to Morton (1970) this effect was due to the fact that tannins destroyed proteins both in the mucosa and in enzymes. Additional *in vitro* data as well as controlled animal studies are necessary to understand how the levels and composition of polyphenols in sorghum affect cancer, and which specific components are responsible. Whole grain consumption has long been correlated with reduced risks to other forms of digestive tract cancer, especially colon cancer. How much of these effects are due to the dietary fibre or the phytochemicals concentrated in the bran of the grains is still unknown. There is strong evidence that phytochemicals present in sorghum have health benefits to humans. Experiments conducted on animals such as rats and guinea pigs have proved these benefits. However, some of the phytochemicals such as condensed tannins, which are beneficial to human health have a negative impact on the productivity of broiler and layer chickens. Therefore, it is safe to assume that sorghum is not generally regarded as an ingredient for controlling poultry health and as such its use has not received significant attention. In this regard, investigations on the use and health benefits of phytochemicals present in sorghum grains are encouraged.

4.3. The Health Benefits of Sorghum Grains for Pigs and Other Animals

Sorghum phytochemicals have been reported to reduce risks associated with cardiovascular disease (CVD) in animals, specifically guinea pigs. This is associated with the consumption of whole grain sorghum which significantly lowers mortality from CVD (Kushi*et al.*, 1999; Slavin*et al.*, 2000; Anderson, 2003). Components of whole grain such as polyphenols and fibre play a role in prevention of CVD. The effects of low-tannin sorghum fed to guinea pigs at 58 % of diet were reported to include cholesterol reduction (Klopfenstein *et al.*, 1981, cited by Awika *et al.*, 2004). More recently, Cho *et al.* (2000) found that sorghum and proso millet hexane extracts inhibit rat liver microsomal 3-hydroxy-3-methylglutaryl CoA (HMGCoA) reductase in a dose-dependent manner. In addition Rooney *et al.* (1992) reported that sorghum and pearl millet brans were excellent bulking agents compared to wheat. They also proved that millet brans have better cholesterol-lowering properties than bran from wheat or white and brown sorghum. The importance of sorghum to cardiovascular health is not limited to positive effects on cholesterol only. Lee and Pan (2003) demonstrated that tannin-sorghum distillery residues inhibited 63-97 % of haemoglobin-catalysed oxidation of linoleic acid in cultured mullet fish. These residues significantly improved blood-thinning and erythrocyte of fish blood cells.The effects are largely attributed to the tannins and other polyphenols present in sorghum grains. There is an association between high plasma cholesterol levels and increased CVD risks although there are no epidemiological data on tannin sorghum consumption and CVD in humans.

Table 2. Antioxidant activity (ORAC) of sorghum grain and bran compared to common fruits

Commodity	ORAC (μmol TE/g dry wt)	References
Tannin sorghum (grain)	868	1
Tannin sorghum (bran)	3124	1
Black sorghum (grain)	219	1
Black sorghum (bran)	1008	1
Red sorghum (grain)	140	1
Red sorghum (bran)	710	1
White sorghum (grain)	22	1
White sorghum (bran)	64	1
Blueberry, lowbush	842	2
Strawberry	402	2
Plum	495	2
Watermelon	18	2
Apple, red delicious	295	2
Orange, navel	135	2

References: 1. Awika *et al.* (2003b). 2. Wu et al. (2004).

There is no direct evidence that sorghum-based diets can improve CVD in pigs. However, based on the study conducted by Klopfenstein *et al.* (1981) on guinea pigs it can be assumed that sorghum diets can promote cardiovascular health in pigs. Investigations on the health benefits of sorghum grains in pigs are encouraged in order to confirm or dispel such a hypothesis.

5. Enhancing the Biomedical Value of Sorghum Grain for Non-Ruminants Animals

5.1. Enhancing Nutritional Value and Utilization of Sorghum Grains for Non-Ruminants Animals

Physical and chemical methods of improving utilization and nutritional value of sorghum grains for non-ruminants animals have been proposed (Babour*et al.*, 2001; Farran*et al.*, 2001). These methods inactivate the anti-nutritional factors in non-conventional feeds in order to improve the digestibility and performance of animals. The physical methods are briefly summarized below:

- Cooking: Mbajuwa (1995) has reported on the effectiveness of reducing anti-nutritional factors (tannin) during the process of cooking, thereby enhancing the nutritional value of the feedstuff.
- Soaking: 24 hours of soaking duration has been shown to improve the nutritional value of cereals and legumes (Ahamefale and Odemelan, 2008).
- Germination: This method has been documented to be an effective treatment to remove/reduce anti-nutritional factors in feedstuffs. Awika and Rooney (2004) showed a decrease in extractable proanthocyanidins, condensed tannins, which are located in the pericarp and testa after germination.

The chemical methods are outlined as follows:

- Use of wood ash: Wood ash is an alkaline substance that is used to reduce tannins in red, bird-resistant sorghum (Etiegni and Campbell, 1991). Such an application of ash improves the digestibility and performance of poultry.
- Addition of fat: Douglas *et al.* (1988) reported that the apparent metabolisable energy (AME) values of low- and high-tannin sorghum were 13.4 and 11.9 mg/kg, respectively and observed that the nutritional value of high-tannin sorghum can be improved if its AME value is increased by the addition of animal fat.
- Use of microbial enzymes: The effectiveness of enzymes to enhance nutrient digestibility has been reported (Choct, 2006). It is well documented that phytase improves phosphorus utilization in poultry (Oyango *et al.*, 2005; Ravindran *et al.*, 2006). Supplementation of feeds for broiler chickens with phytase has been shown to improve growth and bone mineralization, and decreases mortality (Qian *et al.*, 1996).

5.2. Enhancing the Health Benefits of Sorghum Grains for Non-Ruminants Animals

Consumption of sorghum grain has been linked with decreased risks to disease in animals and humans. Therefore, there is a need to develop means to enrich sorghum-based diets with phenolic antioxidants, which have consistent phytochemical profiles. Most phytochemicals in sorghum are concentrated in bran fractions. These fractions can be easily separated from sorghum by decortications, and then be used as dietary supplements. Phytosterols and policosanons are found in sorghum bran and spent distiller's grain in relatively large quantities; these sorghum fractions may be low-cost sources of these valuable compounds. Genetic improvement of sorghum will be a useful strategy to developing cultivars with high yields and higher concentrations of phytochemicals. Sorghum has the potential to be bred specifically to produce high levels of different phenols that can be easily concentrated by simple processes. These special sorghums have reasonable grain yields and agronomic characteristics that make them productive and economical to produce. However, it is important to note that sorghum tannins at high concentrations can reduce weight gain in non-ruminant animals.

Conclusion

Sorghum fractions possess high antioxidant activity *in vitro* relative to other cereals and fruits. These fractions may offer similar health benefits commonly associated with fruits. Information on how sorghum phytochemicals affect poultry and pigs is scarce. However, available epidemiological evidence suggests that consumption of sorghum grain reduces the risks of chronic disease in humans and possibly non-ruminant animals. The high concentration of sorghum phytochemicals is partly responsible. Sorghum phytochemicals also promote cardiovascular health in guinea pigs. It is undeniably important and essential to determine if the positive effects observed *in vitro,* in other animals and in humans can be reproduced in poultry and pigs.

References

Ahamefule, F. O. & Odemelam, V. U. (2008c). Effect of soaking duration on the proximate composition, gross energy, mineral content and some antinutritional properties of *Canavaliaplagiosperma* seed. Proceedings of the 13th Annual Conference of the Animal Science Association of Nigeria (ASAN), 15th-19th Sept., 2008, ABU, Zaria, Nigeria, pp: 491-494.

Aletor, V. A. & Aladetimi, O. O. (1995). Compositional studies on edible tropical species of mushrooms. *Food Chemistry, 54*, 265-268.

Al-Mamary, M., Al-habori, M., Al-Aghbari, A. & Al-Obeidi, A. (2001). In vivo effects of dietary sorghum tannins on rabbit digestive enzymes and mineral absorption. *Nutrition Research*, *21*, 1393–1401.

Ambula, M. K., Oduho,G. W. & Tuitoek, J. K. (2003). Effect of high tannin and bentonite on the performance of laying hens. *Tropical Animal Health and Production*, *35*, 285-292.

Ambula, M. K., Oguho, G. W. & Tuitoek, J. K. (2001). Effects of sorghum tannins, a tannin binder (polyvinylpyrrolidone) and sorghum inclusion level on the performance of broiler chicks. Asian-Australian *Journal of Animal Science*, *14*, 1276–1281.

Anderson, J. W. (2003). Whole grains protect against atherosclerotic cardiovascular disease. *Proceedings of the Nutrition Society*, *62*, 135-142.

Awika, J. M. & Rooney, L. W. (2004). Sorghum phytochemicals and their potential impact on human health. *Phytochemistry*, 65, 1199-1221.

Awika, J. M., McDonough, C. M. & Rooney, L. W. (2005). Decorticating sorghum to concentrate healthy phytochemicals. *Journal of Agricultural and Food Chemistry*, 53, 6230-6234.

Awika, J. M., Rooney, L. W. & Waniska, R. D. (2004). Properties of 3-deoxyanthocyanins from sorghum. *Journal of Agricultural and Food Chemistry*, *52*, 4388–4394.

Awika, J. M., Rooney, L. W., Wu, X., Prior, R. L. & Cisneros-Zevallos, L. (2003). Screening methods to measureantioxidant activity of sorghum (*Sorghum bicolor*) and sorghum products. *Journal of Agricultural and Food Chemistry*, *51*, 6657-6662.

Awika, J. M., Rooney, L. W., Wu, X., Prior, R. L. & Cisneros-Zevallos, L. (2003b). Screening methods to measure antioxidant activity of sorghum (*Sorghum bicolor*) and sorghum products. *Journal of Agricultural and Food Chemistry*, *51*, 6657–6662.

Babour, E. K., Kallas, M. & Farran, M. T. (2001). Immune response to Newcastle disease Virus in broiler chickens: A useful model for the assessment of detoxification of ervil seeds. Revue ScientifiqueEt Technique-Office *International Des Epizooties*, *20*, 785-790.

Boren, B. & Waniska, R. D. (1992). Sorghum seed color as an indicator of tannin content. *Journal of Applied Poultry Research*, *1*, 117-121.

Burdette, A. L. (2003). Nutraceutucaluse of sorghum bran (*Sorghum bicolor*). PhD dissertation, University of Georgia, Athens, USA.

Butler, L. G. (1992). Antinutritional effects of condensed and hydrolysable tannins. In:Plant polyphenols: synthesis, properties and significance (ed. R.W. Hemingway and P.E. Laks), Plenum Press, New York, USA.

Cho, S. H., Choi, Y. & Ha, T. Y. (2000). *In vitro* and *in vivo* effects of prosomillet, buckwheat and sorghum on cholesterol metabolism. *The Journal of the Federation of American Societies for Experimental Biology, 14*, A249.

Choct, M. (2006). Enzymes for the feed industry: Past, present and future. *World's Poultry Science Journal, 62*, 5-15.

Clifford, M. N. (2000). Anthocyanins nature occurrence and dietary burden. *Journal of Science, Food and Agriculture, 80*, 1063-1072.

Cousins, B. W., Tanksley, T. D., Knabe, D. A. & Zebrowska, T. (1981). Nutrient digestibility and performance of pigs fed sorghum varying in tannin concentration. *Journal of Animal Science, 53*, 1524-1529.

Dicko, M. H., Gruppen H., Traore, A. S., Van Berkel, W. J. H. & Voragen, A. G. J. (2005). Evaluation of the effect of germination on phenolic compounds and antioxidant activities in sorghum varieties. *Journal of Agricultural and Food Chemistry, 53*, 2581-2588.

Douglas, J. H., Sullivan, T. W., Bond P. I. & Rydeli, J. A. (1988). Use of animal fat to correct the lowermetabolizable energy (ME) and nutritional value of high-tannin sorghum grain. *Poultry Science, 67*, 808-901.

Dowling, L. F., Arndt, C. & Hamaker, B. R. (2002). Economic viability of high digestibility sorghum as feed for market broilers. *Agronomy Journal, 94*, 1050-1058.

Dykes, L. & Rooney, L. W. (2006). Sorghum and millet phenols and antioxidants. *Journal of Cereal Science, 44*, 236-251.

Dykes, L., Rooney, L. W., Waniska, R. D. & Rooney, W. L. (2005). Phenolic compounds and antioxidant activity of sorghum grains of varying genotypes. *Journal of Agricultural and Food Chemistry, 53*, 6813-6818.

Dykes, L., Rooney, L. W., Waniska, R. D. & Rooney, W. L. Phenolic compounds and antioxidants activity of sorghum grains of varying genotypes. *Journal of Agricultural and Food Chemistry, 53*, 6813.

Etiegni, L. & Campbell, A. G. (1991). Physical and chemical characteristics of wood ash. *Bioresource Technology Journal, 37*, 173-178.

Etuk, E. B. (2008). Nutrient composition and feeding value of sorghum for turkeys. PhD Thesis, Federal University of Technology, Owerri, Nigeria.

Farran, M. T., Dakessian, P. B., Darwish, A. H., Uwayjan, M. G., Dbouk, H. K., Seliman, F. T. & Ashkarian, V. M. (2001). Performance of broiler chicken production and egg quality parameters of layinghens fed 60 % raw and treated common vetch (*Viciaervilia*) seeds. *Poultry Science Journal, 80*, 203-208.

Featherson, W. R. & Rogler, J. C. (1975). Influence of tannin on the utilization of sorghum grain by rats and chickens. *Nutrition Reports International*, *11*, 491-497.

Gomez-Cordoves, C., Bartolome, B., Vieira, W. & Virador, V. M. (2001). Effects of wine phenolics and sorghum tannins on tyrosinase activity and growth of melanoma cells. *Journal of Agricultural and Food Chemistry*, *49*, 1620–1624.

Gous, F. (1989). Tannins and phenols in black sorghum. Ph.Ddissertation, Texas A&M University, College Station, Texas, USA.

Hagerman, A. E. & Butler, L. G. (1981). The specificity of proanthocyanidin–protein interactions. *Journal of Biological Chemistry*, *256*, 4494-4497.

Hagerman, A. E., Riedl, K. M., Alexander Jones, G.,Sovik, K. N., Ritchard, N. T., Hartzfeld, P. W. & Riechel, T. L. (1998). High molecular weight plant polyphenolics (tannins) as biological antioxidants. *Journal of Agricultural and Food Chemistry*, *46*, 1887-1892.

Hahn, D. H., Rooney, L. W. & Faubion, J. M. (1983). Sorghum phenolic acids, their HPLC separation and their relation to fungal resistance. *Cereal Chemistry*, *60*, 255-259.

Haslam, E. (1974). Polyphenol-protein interactions. *Biochemical Journal*, *139*, 285-288.

Holtekjølen, A. K., Kinitz, C. & Knutsen, S. H. (2006). Flavanol and bound phenolic acid contents in different barley varieties. *Journal of Agricultural and Food Chemistry*, *54*, 2253-2260.

Hu, C., Zawistowski, J., Ling, W. & Kitts, D. D. (2003). Black rice (*Oryzasativa L. indica*) pigmented fraction suppresses both reactive oxygen species and nitric oxide in chemical and biological model systems. *Journal of Agricultural and Food Chemistry*, *51*, 5271-5277.

Jacob, J. P., Mitaru, B. N., Mbugua, P. N. & Blair, R. (1996). The effects of substituting Kenyan Serena sorghum for maize in broiler starter diets with different dietary crude protein and methionine levels. *Animal Feed Science and Technology*, *61*, 27-39.

Jambubathan, R. & Mertz, E. T. (1973). Relationship between tannin levels, rat growth, and distribution of proteins in sorghum. *Journal of Agricultural and Food Chemistry*, *21*, 692-696.

Kahlon, T. S. & Shao, Q. (2004). *In vitro* binding of bile acids by soybean (Glycine max), black eye bean (*Vignaunguiculata*), garbanzo(*Cicerarietinum*) and lima bean (*Phaseoluslunatus*). *Food Chemistry*, *86*, 435-440.

Kim, K. H., Tsao, R., Yang, R. & Cui, S. W. (2006). Phenolic acid profiles and antioxidant activities of wheat bran extracts and the effect of hydrolysis conditions. *Food Chemistry*, *95*, 466-473.

Klopfenstein, C. F., Varriano-Marston, E. & Hoseney, R. C. (1981). Cholesterol-lowering effect of sorghum diet in guinea pigs. *Nutrition Reports International*, *24*, 621-626.

Kushi, L. H., Meyer, K. A. & Jacobs, D. R. (1999). Cereals, legumes, and chronic disease risk reduction: evidence from epidemiologic studies. *American Journal of Clinical Nutrition*, *70* (Suppl.), 451S-458S.

Lee, S. M. & Pan, B. S. (2003). Effect of dietary sorghum distillery residue on hematological characteristics of cultured grey mullet (*Mugilcephalus*)—an animal model for prescreening antioxidant and blood thinning activities. *Journal of Food Biochemistry*, *27*, 1-18.

Liu, R. H. (2004). Potential synergy of phytochemicals in cancer prevention: mechanism of action. *Journal of Nutrition*, *134*, 3479S-3485S.

Lizardo, R., Peiniau, J. & Aumaitre, A. (1995). Effect of sorghum onperformance, digestibility of dietary-components and activities ofpancreatic and intestinal enzymes in the weaned piglet. *AnimalFeed Science and Technology*, *56*, 67-82.

Mathanghi, S. K. (2012). Nutraceutical properties of great millet-Sorghum Vulgaris. Internation Journal of Food, *Agriculture and Veterinary Sciences*, *2*, 40-45.

Mattila, P., Hellstrom, J. & Torronen, R. (2006). Phenolicc Acids in berries, fruits and vegetables. *Journal of Agricultural and Food Chemistry*, *54*, 7193–7199.

Mattila, P., Pihlava, J. M. & Hellström, J. (2005). Contents of phenolic acids, alkyl- and alkenylresorcinols, and avenanthramides in commercial grain products. *Journal of Agricultural and Food Chemistry*, *53*, 8290-8295.

Mbajuwa, O. K. (1995). Effect of processing on some antinutritive and toxic components of the African oil beanseed (*Pentaclethramacrophyla*Benth). *Journal of Food Science and Agriculture*, *68*, 153-158.

McDonough, C. M., Rooney, L. W. & Serna-Saldivar, S. O. The millets.In:The handbook of cereal science and technology. Kulp, K. and Ponte, J.G..JredsMarce Dekker, Inc., New York, USA.

Morton, J. F. (1970). Tentative correlations of plant usage and esophageal cancer zones. *Economic Botany*, *24*, 217-226.

Muriu, J. I., Njoka-Njiru, E. N., Tuitoek, J. K. & Nanua, J. N. (2002). Evaluation of sorghum (*Sorghum bicolor*) as replacement for maize in the diet of growing rabbits (*Oryctolaguscuniculus*). *Asian-Australian Journal of Animal Science*, *15*, 565-569.

Naczk, M. & Shahidi, F. (1997). Nutritional implications of canola condensed tannins. Antinutrients and chemicals in food: ACS *Symposium Series*, *662*, 186-208.

National Research Council. (1984). Nutrient Requirements of Poultry, 9th ed. National Academy of Science, Washington DC, USA.

Neucere, N. J. & andSumrell, G. (1980). Chemical composition of different varieties of grain sorghum. *Journal of Agricultural and Food Chemistry*, *28*,19-21.

Nguz. K., Van Gaver, D. & Huyghebaert, A. (1998). *In vitro* inhibition of digestive enzymes by sorghum condensed tannins (*Sorghum bicolor* L. (Moench)). *Science Des Aliment*, *18*, 507-514.

Olomu, J. M. (1995). Monogastricanimal nutrition: Principles and practice. *A Jackem Publication Benin City*, (1st Ed.), pp: 285.

Oyango, E. M., Bedford, M. R. & Adeola, O. (2005). Efficacy of an evolved *Escherichia coli* phytase in diets of broiler chicks. *Poultry Science Journal*, *84*, 248-255.

Oyenuga, V. A. (1968). *Nigeria's food and feeding stuff – Their chemistry and nutritive value*. Ibadan University Press, Ibadan, Nigeria, 39-41.

Parr, A. J. & Bolwell, G. P. (2000). The potential for possible nutritional enhancement of the diet by modifying the phenols content or profile. *Journal of Food Science and Agriculture*, *80*, 985-1012.

Purseglove, J. W. (1972). Tropical Crops -Monocotyledons, Longman Group Ltd., UK, 261–286.

Ravindran, V., Morel, P. C. H., Partridge, G. G., Hruby, M. & Sands, J. S. (2006). Influence of a Escherichia Coli-derived phytase on nutrients utilization in broiler starters fed diets containing varying concentration of phytic acid. *Poultry Science*, *85*, 82-89.

Rooney, T. K., Rooney, L. W. & Lupton, J. R. (1992). Physiological characteristics of sorghum and millet brans in the rat model. *Cereal Foods World*, *37*, 782–786.

Sedghi, M., Ebadi, M. R., Golian, A. & Ahmadi, H. (2011). Estimation and modeling true metabolisableenergy of sorghum grain for poultry. *Poultry Science*, *90*, 1138–1143.

Slavin, J. (2004). Whole grains and human health. *Nutrition Research Reviews.*, *17*, 99-110.

Slavin, J. L., Jacobs, D. & Marquart, L. (2000). Grain processing and nutrition. *Critical Reviews in Food Science and Nutrition*, *40*, 309-326.

Tacon, A. G. J. (1995). *Semi-intensive feeding methods for freshwater fish: concepts and research approaches*. In: The management of integrated freshwater agro-piscicultural ecosystems in tropical areas. Seminar, Brussels, 16-19, May 1994.

Walker, T. (1999). Physical aspects of grain and effect of feed texture on animal performance. ASA *Technical Bulletin*, AN22-1999.

Waniska, R. D. (2000). Structure, phenolic compounds and antifungal proteins of sorghum caryopses, pp. 72–106. In: A., Chandrashekar, R., Bandyopadhyay, A. J. Hall, eds. Technical and institutional options for sorghum grain mold management: proceedings of an international consultation, ICRISAT.

Waniska, R. D., Poe, J. H. & Bandyopadhyay, R. (1989). Effects of growth conditions on grain molding and phenols in sorghum caryopsis. *Journal of Cereal Science*, *10*, 217–225.

Wu, Y. B. & Ravindran, V. (2004). Influence of whole-wheat inclusion and xylanase supplementation on the performance, digestive tract measurements and carcass characteristics of broiler chickens. *Animal Feed Science and Technology*, *116*, 129-139.

Yao, L. H., Jiang, Y. M., Shi, J., Tomas-Barbéran, F. A., Datta, N., Singanusong, R. & Chen, S. S. (2004). Flavonoids in food and their health benefits. *Plant Foods Human and Nutrition*, *59*, 113-122.

In: Sorghum
Editor: Patricia C. Parra

ISBN: 978-1-62618-351-3

Chapter 4

FUNGAL ROOT ENDOPHYTES AS SORGHUM GROWTH PROMOTERS

***A. B. M. Vaz*[1,2,*], *I. Sampedro*[3,†] *and J. A. Ocampo*[3,‡]**

[1]Research Laboratory in Microbiology (LAPEM), Biological Sciences Department, Feira de Santana State University, Feira de Santana, Bahia, Brazil

[2]Center for Excellence in Bioinformatics, Genomics and Computational Biology Group. Centro de Pesquisas René Rachou (CPqRR) – FIOCRUZ, Minas Gerais, Brazil

[3]Department of Microbiology, Estación Experimental del Zaidín, C.S.I.C., Profesor Albareda, Granada, Spain

ABSTRACT

Fungal endophytes are found in healthy plant tissues during at least one stage of their life cycle and do not cause any apparent symptoms of disease or negatively affect the host. These fungi have been isolated from

* Corresponding author: Aline B. M. Vaz. 1. PPG Biotecnologia UEFS/Fiocruz-BA. Av. Transnordestina s.n. Novo Horizonte, 44.036-900, Feira de Santana, BA – Brasil. 2. Centro de Excelência em Bioinformática (CEBio), Fundação Oswaldo Cruz (FIOCRUZ), Rua Araguari 741, 30190-110 Belo Horizonte, MG, Brazil. E-mail: alinebmv@hotmail.com.

† I. Sampedro: Department of Microbiology, Estación Experimental del Zaidín, C.S.I.C., Profesor Albareda, 1, E-18008, Granada, Spain.

‡ J. A. Ocampo: Department of Microbiology, Estación Experimental del Zaidín, C.S.I.C., Profesor Albareda, 1, E-18008, Granada, Spain. E-mail: juanantonio.ocampo@eez.csic.es.

all plants studied in natural ecosystems. They can induce many changes in the plant ecology, fitness and evolution and, thus, can influence whole plant communities. The Ascomycetous root fungal endophytes can be divided into two groups, the dark septate endophytes (DSE) and the fungi with hyaline and pale hyphae. Fungal endophyte colonization is important for the improvement of the ecological adaptability of the host, resulting in enhanced tolerance of both biotic and abiotic stresses. Another important group of endophytes are those that constitute arbuscular mycorrhizal (AM) symbiosis. AM symbiosis is very important for plant nutrient uptake and growth and for plant resistance to biotic and abiotic stresses. It is known that AM symbiosis can be influenced by other root fungal endophytes, which confer benefits to the host plant, such as growth promotion, protection against disease or enhanced phosphorous uptake. Despite the benefits of these root fungal endophytes for agronomical plants, the physiological and ecological significance of these fungi remains poorly characterized. In this context, we discuss the potential roles of fungal endophytes for growth promotion in *Sorghum*.

Fungal Endophyte Diversity

Fungal endophytes are microorganisms that are present in healthy plant tissues during at least one stage of their life cycle and do not cause any disease or exert negative effects on their hosts (Petrini *et al.*, 1992). Fungal endophytes live entirely within the plant tissues and may grow within the roots, stems and leaves, emerging only to sporulate during plant or host-tissue senescence, unlike mycorrhizal fungi that colonize plant roots and grow into the rhizosphere (Stone *et al.*, 2004). Endophytes have been isolated from all previously studied plants, including bryophytes (U'Ren *et al.*, 2010), pteridophytes (Petrini *et al.*, 1992), gymnosperms (Soca-Chafre *et al.*, 2011), and both monocotyledonous (Pinruan *et al.*, 2010) and dicotyledonous angiosperms (Vaz *et al.*, 2009). These fungi colonize all available tissues, such as roots, stems, leaves, bark, fruits, seeds, and floral organs (Petrini *et al.*, 1992; Rodriguez *et al.*, 2009; Vaz *et al.*, 2009).

According to Rodriguez *et al.* (2009), fungal endophytes can be divided into Clavicipitaceous (C-endophytes, class 1) and Nonclavicipitaceous (NC-endophytes, classes 2 to 4). The C-endophytes infect some grasses and grow systematically within aerial tissues; their transmission is primarily vertical (Saikkonen *et al.*, 2002; Clay and Schardl, 2002). Association with C-endophytes results in an increase in the plant biomass and drought tolerance and in the production of a diverse array of secondary metabolites that confer protection against herbivores and insects (Clay, 1988; Schulz and Boyle, 2005). The benefits for the fungal partner are not only nutritional; they also include protection from abiotic stresses (Bacon and Hill, 1996). The NC-endophytes have been recovered from every major lineage of land plants and from all terrestrial ecosystems. They can be differentiated into three functional classes (Classes 2 to 4, as proposed by Rodriguez *et al.*, 2009) based on their host colonization patterns, mechanisms of transmission between host generations, *in planta* biodiversity levels and ecological functions (Rodriguez *et al.*, 2009). Most of these fungi belong to the Ascomycota (Petrini *et al.*, 1986), colonize inter- or intracellularly and result in localized or systemic infections (Schulz and Boyle, 2005). Very little is known regarding the costs and benefits of endophyte infection. Some studies have shown that the NC-endophytes establish mutualistic relationships with plants and that these relationships confer fitness benefits, such as biotic and abiotic stress tolerance, nutrient acquisition and increased growth and yields, to both plant and fungus (Rodriguez *et al.*, 2009). Root fungal endophytes belong mainly to Class 2 and Class 4. Root colonization by fungal endophytes may confer benefits to the host plant including growth promotion, protection against disease, enhanced tolerance to biotic and abiotic stresses and facilitation of phosphorous uptake (Sieber, 2002; Schulz and Boyle, 2005). Moreover, some endophytes are able to enhance the growth of many plant species without the need for concomitant nutrient uptake (Kageyama *et al.*, 2008).

ROOT FUNGAL ENDOPHYTES

Class 2 and Class 4

Most fungal endophytes belonging to Class 2 also belong to Ascomycota or Pezizomycotina, with a minority belonging to Basidiomycota, Agaricomycotina and Pucciniomycotina. These fungi are known for their hyaline and pale hyphae and are less studied because they are less conspicuous and easier to overlook than the DSE (Addy *et al.*, 2005). However, these fungal endophytes are known to be able to interact with many host plants and provide many beneficial effects to their host partner by increasing the plant biomass (Gasoni and Gurfunkel, 1997), acting as a biological control agent (Maciá-Vicente *et al.*, 2008) and conferring protection against abiotic and biotic stresses (Rodriguez *et al.*, 2008).

Many reports show that the fungal endophytes isolated from natural vegetation can colonize the rhizosphere of non-host plants without causing any disease symptoms. As hypothesized by Rodriguez *et al.* (2008), the fungal endophytes confer protection in a habitat-specific manner that can be explained by an intergenomic epigenetic mechanism for plant adaptation to habitat stresses. These studies have shown that the class 2 endophytes can have significant effects on the ecophysiology of plants in addition to their potential benefits to important agricultural crop species.

Class 4 is comprised of the DSEs. These fungal endophytes are mainly sterile septate fungi with melanised hyphae and microesclerotia that colonize the inter- and the intracellular root tissue (Jumpponen and Trappe, 1998; Addy *et al.*, 2005). The most common DSEs belong to the Leotiales, Sordariales (Grünig *et al.*, 2002) and Chaetothryales (Narisawa *et al.*, 2007). The ascomycetous DSEs likely constitute the most abundant and most widespread group of root colonizers and are similar to the arbuscular mycorrhizal (AM) fungi with regard to the occurrence and colonization of plant species; these two fungal groups coexist in the soil (Mandyam and Jumpponen, 2005; Summerbell, 2005). The fungal endophytes colonize the root surface and frequently form a network of mycelium over the surface of and within the root (Peterson *et al.*, 2004). Fungal hyphae enter the host through root hairs, epidermal cells, root cap cells if the root apical meristems are invaded, or through wounds (Peterson *et al.*, 2004).

The majority of structural studies on DSEs have involved *Phialocephala fortinii* Wang and Wilcox because of its widespread geographical distribution, its occurrence in various ecosystems, its broad host range and the ease with which it can be cultured (Peterson *et al.*, 2008). By observing the root colonization of these fungi, it is possible to identify surface hyphae, intracellular hyphae, microscletoria (Peterson *et al.*, 2008) a mantle (Wilcox and Wang, 1987) and a Hartig net (Fernando and Currah, 1996). Narrow hyphae are able to colonize the epidermal cells without distorting the cell wall, which is suggestive of an enzymatic process (Currah *et al.*, 1993). Moreover, the spread of colonization from epidermal cell to epidermal cell is achieved via narrow penetration hyphae (Currah *et al.*, 1993). Root fungal endophyte colonization depends on plant species and the environment in which colonization occurs (Peterson *et al.*, 2008). However, by observing the colonization pattern of the fungal endophytes, it is not possible to determine whether they are biotrophics, hemibiotrophics, necrotrophics, or simply commensals. More detailed structural studies are needed to determine the effect of these fungi on the plant's growth (Peterson *et al.*, 2008).

The mechanisms underlying the beneficial interaction between DSEs and their hosts have remained elusive. Mandyam and Jumpponen (2005) suggested that the fungal endophyte hyphae penetrate into the soil micropores where the plant root cannot reach and thus provide the plant access to soluble nutrient sources that were otherwise unavailable. Moreover, DSEs may improve plant growth by facilitating the use of organic nutrient sources by means of their hydrolytic enzymes (Mandyam and Jumponen, 2005).

Many studies have isolated DSE fungal endophytes and reinoculated them onto the same species or in another species to observe any resulting growth effects. These studies showed that some fungal endophytes are able to increase plant biomass (Jumpponen and Trape, 1998; Newsham, 1999; Usuki and Narisawa, 2007). Moreover, some studies have indicated that these fungi are able to increase the N and P uptake (Jumpponen and Trape, 1998, Newsham, 1999, Usuki and Narisawa, 2007).

ARBUSCULAR MYCORRHIZAS (AM)

The AM fungi are the most widespread root endophyte-associated fungi. However, they are considered separately because of their specificity in their associations, the formation of specialized structures, the synchronized development between the fungi and the host and the significant benefits for both partners (Brundrett, 2006). The AM fungi belong to the Glomeromycota (Oehl *et al.*, 2011; Schüßler and Walker, 2010). AM are considered to be obligate biotrophs because they are unable to grow and complete their life cycle in the absence of an association with a living root (Smith and Read, 2008). The fungus lives partially inside the root, forming internal hyphae, arbuscules and vesicules, and partially outside the root, forming an external mycelium, BAS (branched absorbing structures), auxiliary cells and spores (Smith and Read, 2008). This symbiosis can benefit the plant growth, primarily by increasing the capability of the root system to absorb and translocate phosphorous from the soil through an extensive external hyphae network. Additionally, the fungus obtains carbohydrates from the photosynthesizing plant via the arbuscules (the plant cell plasmalemma interface with the fungus). This symbiosis also provides a protected site within the root cells where the fungus can live (Morton *et al.* 2004, Smith and Read, 2008). Sorghum has historically been used to develop soil-root inoculums and as a suitable host for the maintenance and mass production of AM fungi (Bagyaraj and Manjunath, 1980; Kavatagi and Lakshman, 2012). Research on sorghum-AM symbiosis began decades ago, and the sorghum P uptake via AM fungi has been the subject of research for many years (Ratnayake *et al.*, 1978; Bethlenfalvay *et al.*, 1985; Hata *et al.*, 2010; Calderón *et al.*, 2012; Whiteside *et al.*, 2012). Many studies have investigated the beneficial effects of the AM fungi on the survival and growth of sorghum. Moreover, studies on the uptake of neutral and positively charged amino acids (N and P) by sorghum plants grown under greenhouse and field conditions have also been performed (Fracchia *et al.*, 2000; Whiteside *et al.*, 2012). The AM colonization of sorghum improves the ability of sorghum to compete with other plants for soil nutrients (Ocampo, 1986). The role of the AM pathway during P uptake has also been identified as important for the enhanced ability of these plants to compete with their non-mycorrhizal neighbors (Smith and Smith, 2012). Nutrient transfer among sorghum plants and between sorghum and other plants, possibly through the AM mycelial connections between the roots, seems to be implicated in sorghum's ability to compete (Ocampo, 1986). Finally, the AM fungal hyphae increase the soil aggregation capacity of the sorghum roots (Siddiky *et al.*, 2012).

Because the AM fungi are partially inside and partially outside of the host, external factors, such as herbicides and drought, will affect the development of symbiosis. The application of herbicides negatively affected the colonization and metabolic activity of the AM fungi in the sorghum roots (Ocampo and Barea, 1985). However, the AM fungi conferred resistance against the negative effects of herbicides to the sorghum plants (Ocampo and Barea, 1985; Mujica *et al.*, 1999). Additionally, AM fungi alleviated the negative effects of drought stress on the sorghum plant's length, shoot dry matter and grain yield (Alizadeh *et al.*, 2011). The contribution of AM fungi to the regulation of the stomatal behavior in sorghum plants can be implicated in the plants' resistance to drought stress (Augé *et al.*, 2007).

Fungal Root Endophytes and Sorghum Growth

Few studies have considered the use of class 2 and class 4 endophytes (Rodriguez *et al.*, 2009) as growth promoters of *Sorghum vulgare*. Experiments on the effect of fungal endophytes isolated from *Myrtus communis* on the growth of *S. vulgare* have been performed previously (Vaz *et al.*, 2012). These fungal endophytes were identified by sequencing the internal transcribed spacer (ITS1 and ITS2) and 5.8S regions of the nuclear ribosomal repeat unit and were identified as *Phomopsis columnaris*, *Cylindrocarpon destructans*, *Bionectria ochroleuca* 1, *Bionectria ochroleuca* 2, *Phoma schachtii*, *Eucasphaeria* sp., *Cylindrocarpon pauciseptatum* and *Bisporella* sp. Although none of these fungal endophytes were able to colonize the sorghum roots, all of them exerted effects on their host partner.

The fungal endophytes *C. destructans* and *P. schachtti* increased the shoot dry matter values of the sorghum plants grown on sterile soil. Some root fungal endophytes are known to act as saprobe fungi and occur in the rhizosphere to obtain greater nutritional benefit from the organic and inorganic compounds released from the living roots and the sloughed cells (Scervino *et al.*, 2009). The *Helminthosporium velutinum* Link ex Ficinus and Schubert fungal endophyte was isolated from sweet Sorghum roots and significantly increased the sorghum biomass relative to other fungal endophytes, including the frequently reported representatives of the DSEs, such as *Cadophora finlandica*, *Heteroconium chaetospira* and *Phialocephala fortinii* (Diene *et al.*, 2010).

The interactions of plants with endophytes or mycorrhizal fungi have been studied extensively; however, studies about tripartite interactions between endophytes, mycorrhizal fungi and plants are scarce (Reininger and Sieber, 2012). The fungal endophytes *Phomopsis columnaris*, *Phoma schachtii*, *Eucasphaeria* sp., and *Cylindrocarpon pauciseptatum* behave similar to saprobe fungi and exert synergistic effects on the plant dry matter and on AM colonization (Vaz *et al.*, 2012). The root tissue, rhizosphere and soil form a complex morphological, physical and chemical microcosm that provides a broad range of different habitats in which a myriad of different microorganisms are able to colonize, grow and influence each other's growth. Several mechanisms implicated in the interactions between AM and saprobe fungi have been described previously (Fracchia *et al.*, 2000; Arriagada *et al.*, 2010). Some soil microorganisms are able to mobilize and promote more efficient nutrient by AM-colonized plants (Fracchia *et al.*, 2000). Furthermore, some soil microorganisms can produce substances that stimulate plant growth or inhibit pathogens (Jeffries and Dodd 1996). Mycorrhiza formation can affect the microbial population in the rhizosphere directly or indirectly through changes in the root exudation patterns or through fungal exudates (Linderman 1992). Some DSEs, such as *Drechslera* sp. and other endophytic fungi, are known to be able to colonize the roots of grasses such as sorghum, and the exudates produced by the host can stimulate the development of the presymbiotic stage of AM fungi (Fracchia *et al.*, 2000, 2004, Navarro, 2008; Scervino *et al.*, 2008, 2009). Scervino *et al.* (2009) showed that the exudates produced by the DSE fungus stimulated the hyphal length and branching of the

AM fungus. These authors also proposed that an increase in the presymbiotic stage of the AM fungus could be thought of as an increase in the root colonization.

Under greenhouse and field conditions, the application of *Fusarium oxysporum* increased the shoot dry matter, the N and P concentrations of sorghum plants, and the level of AM colonization achieved by indigenous or introduced AM fungi (Fracchia *et al.*, 2000). The use of organic nutrient sources by the hydrolytic enzymes produced by the DSEs is one of the areas where DSEs may complement the functions of AM fungi (Mandyam and Jumponen, 2005). The increase of the shoot dry matter of the sorghum that was inoculated with *Eucasphaeria* sp. coincided with an increase in the P concentration of the shoot of the AM-colonized sorghum plants. However, both strains of *B. ochroleuca* increased the AM root colonization but not the plant dry matter (Vaz *et al.*, 2012). There are indications that the soil fungi benefit from the AM symbiosis more by increasing the AM fungal development in the plant root than by increasing the plant dry matter through a nutritional effect (Aranda *et al.*, 2007). Nevertheless, *Ph. columnaris*, *P. schachtii*, and *C. pauciseptatum* can increase the plant dry matter but cannot increase the nutrient concentration of the sorghum, suggesting that their beneficial effect on the plant dry matter occurs via other mechanisms, such as phytohormone production.

CONCLUSION

Root fungal endophytes are important for plant physiology and ecology. The interaction between AM fungi and other root fungal endophytes can greatly affect crop productivity and ecosystem sustainability via new plant production strategies. For millions of people in the semi-arid tropics of Asia and Africa, sorghum is one of the most important staple foods. This crop sustains the lives of the poorest rural people and will continue to do so in the foreseeable future. Sorghum grows in harsh environments where other crops do not grow as well. Improvements in the production, availability, storage, utilization and consumption of this food crop will significantly contribute to the household food security and nutrition of the inhabitants of these areas (FAO, 1996). Moreover, there is great interest in using Sorghum for biofuel production as an alternative to non-renewable energy sources (Chandel *et al.*, 2007). Previous studies have shown the positive effects of fungal endophytes on AM colonization and dry matter growth of hosts, reinforcing the potential use of these fungi as growth promoters of sorghum; additionally, the utilization of fungal endophytes could be of financial importance with regard to crop improvement and biofuel production.

REFERENCES

Addy, H. D., Piercey, M. M., Currah, R. S. (2005). Microfungal endophytes in roots. *Canadian Journal of Botany*, 83: *1*–13.

Alizadeh, O., Zare, M., Nasr, A. H. (2011). Evaluation effect of mycorrhiza inoculate under drought stress condition on grain yield of sorghum (*Sorghum bicolor*). *Advances in Environmental Biology*, 5: 2361-2364.

Aranda, E., Sampedro, I., Díaz, R., García, M., Ocampo, J. A., García-Romera, I. (2007). Xyloglucanases in the interaction between saprobe fungi and the arbuscular mycorrhizal fungus *Glomus mosseae*. *Journal of Plant Physiology*, 164: 1019-1027.

Arriagada, C., Pereira, G., García-Romera, I., Ocampo, J. A. (2010). Improved zinc tolerance in *Eucalyptus globulus* inoculated with *Glomus deserticola* and *Trametes versicolor* or *Coriolopsis rigida*. *Soil Biology and Biochemistry,* 42: 118-124.

Augé, R. M., Toler, H. D., Moore, J. L., Cho, K., Saxton, A. M. (2007). Comparing contributions of soil versus root colonization to variations in stomatal behavior and soil drying in mycorrhizal *Sorghum bicolor* and *Cucurbita pepo*. *Journal of Plant Physiology*, 164: 1289-1299.

Bacon, C. W. and Hill, N. S. Symptomless grass endophytes: products of coevolutionary symbiosis and their role in the ecological adaptations of infected grasses. In: Redlin, S. C. and Carris, L. M. (eds). *Endophytic fungi in Grasses and Woody Plants.* St Paul, MN.: American Phytopathological Society Press; 1996, pp 155-178.

Bagyaraj, D. J. and Manjunath, A. (1980). Selection of a suitable host for mass production of VA mycorrhizal inoculum. *Plant and Soil,* 55: 495-498.

Bethlenfalvay, G. J., Ulrich, J. M., Brown, M. S. (1985). Plant response to mycorrhizal fungi: host, endophyte, and soil effects. *Soil Science Society of America Journal*, 49: 1164-1168.

Brundrett, M. C. Understanding the roles of multifunctional mycorrhizal and endophytic fungi. In: Schulz, B., Boyle, C., Sieber, T N. (eds). *Microbial root endophytes*. Berlin: Springer; 2006; pp. 179–190.

Calderón, F. J., Schultz, D. J., Paul, E. A. (2012). Carbon allocation, belowground transfers, and lipid turnover in a plant-microbial association. *Soil Science Society of America Journal*, 76: 1614-1623.

Chandel, A. K., Chan, E. S., Rudravaram, R., Narasu, M. L., Rao, L. V., Ravindra, P. (2007). Economics and environmental impact of bioethanol production technologies: An appraisal. *Biotechnology and Molecular Biology Review,* 2:14–32.

Clay, K. and Schardl, C. L. (2002). Evolutionary origins and ecological consequences of endophyte symbiosis with grasses. *American Naturalist*, 160: S99–S127.

Clay, K. (1988). Fungal endophytes of grasses: a defensive mutualism between plants and fungi. *Ecology*, 69: 10–16.

Currah, R. S., Tsuneda, A., Murakami, S. (1993). Morphology and ecology of *Phialocephala fortinii* in roots of *Rhododendron brachycarpum. Canadian Journal of Botany*, 71: 1639-1644.

Diene, O., Takahashi, T., Yonekura, A., Nitta, Y., Narisawa, K. (2010). A New Fungal Endophyte, *Helminthosporium velutinum*, Promoting Growth of a Bioalcohol Plant, Sweet Sorghum. *Microbes and Environment*, 25: 216–219.

FAO (1996). International Crops Research Institute for the Semi-Arid Tropics (ICRISAT)/Food and Agriculture Organization (FAO). The World Sorghum and Millet Economies. ICRISAT, Patancheru, India/ FAO, Rome.

Fernando, A. A. and Currah, R. S. (1996). A comparative study of the effects of the root endophytes Leptodontidium orchidicola and *Phialocephala fortinii* (Fungi Imperfecti) on the growth of some subalpine plants in culture. *Canadian Journal of Botany*, 74: 1071-1078.

Fracchia, S., García-Romera, I., Godeas, A., Ocampo, J. A. (2000). Effect of the saprophytic fungus *Fusarium oxysporum* on arbuscular mycorrhizal colonization and growth of plants in greenhouse and field trials. *Plant and Soil,* 223: 175-184.

Fracchia, S., Sampedro, I., Scervino, J. M., García-Romera, I., Ocampo, J. A., Godeas, A. (2004). Influence of saprobe fungi and their exudates on arbuscular mycorrhizal symbioses. *Symbiosis*, 36: 169-182.

Gasoni, L. and De Gurfinkel, B. S. (1997. The endophyte *Cladorrhinum foecundissimum* in cotton roots: phosphorus uptake and host growth. *Mycological research*, 101: 867-870.

Grünig, C. R., Sieber, T. N., Holdenrieder, O. (2002). Genetic variability among strains of *Phialocephala fortinii* and phylogenetics analysis of the genus *Phialocephala* based on rDNA ITS sequence comparisons. *Canadian Journal of Botany*, 80: 1239-1249.

Hata, S., Kobae, Y., Banba, M. (2010). Interactions between plants and arbuscular mycorrhizal fungi. *International Review of Cell and Molecular Biology,* 281: 1-48.

Jeffries, P. and Dodd, J. C. Functional ecology of mycorrhizal fungi in sustainable soil-plant systems. In: Azcon-Aguilar, C. and Barea, J. M. (eds). *Mycorrhizas in integrated systems from genes to plant development.* Brussels: European Commission, 1996, pp. 497–501.

Jumpponen, A. and Trappe, J. M. (1998). Dark septate endophytes: a review of facultative biotrophic root-colonizing fungi. *New Phytologist*, 140: 295-310.

Kageyama, S. A., Mandyam, K. G., Jumpponen, A. Diversity, function and potential applications of the root-associated endophytes. In: Varma, A. (ed.). *Mycorrhiza – State of the art, genetics and molecular biology, eco-function, biotechnology, ecophysiology, structure and systemetics*. Berlin: Springer-Verlag, 2008, pp. 29-57.

Kavatagi, P. K. and Lakshman, H. C. (2012). Effect of arbuscular mycorrhizal fungi for their symbiotic efficiency on two varieties of *Solanum lycopersicum* L. *International Journal of Pharma and Bio Sciences*, 3: B1007-B1017.

Linderman, R. G. Vesicular-arbuscular mycorrhizae and soil microbial interactions. In: Bethlenfalvay, G. J. and Linderman, R. G. (Eds.), *Mycorrhizae in Sustainable Agriculture.* Madison, Wisconsin: ASA Spec. Publ., 1992, pp. 45–70.

Maciá-Vicente, J. G., Jansson, H. B., Mendgen, K., Lopez-Llorca, L. V. (2008). Colonization of barley roots by endophytic fungi and their reduction of take-all caused by *Gaeumannomyces graminis* var. *tritici*. *Canadian Journal of Microbiology*, 54: 600-9.

Mandyam, K. and Jumpponen, A. (2005). Seeking the elusive function of the root-colonising dark septate endophytic fungi. *Studies in Mycology*, 53: 173:189.

Morton, J. B., Koske, R. E., Stürmer, S. L., Bentivenga, S. P. Mutualistic arbuscular endomycorrhizal fungi. In: Mueller, G., Bills, G., Foster, M. (eds.). Measuring and Monitoring Biodiversity of Fungi. *Inventory and monitoring methods.* Boston, MA: Elsevier Academic Press, 2004, pp. 317- 336.

Mujica, M. T., Fracchia, S., Ocampo, J. A., Godeas, A. (1999). Influence of the herbicides chlorsulfuron and glyphosate on arbuscular mycorrhizas of

Glycine max intercropped with the weeds *Brassica campestris* and *Sorghum halepensis*. *Symbiosis*, 27: 73-91.

Narisawa, K., Hambleton, S., Currah, R. S. (2007). *Heteroconium chaetospira*, a dark septate root endophyte allied to the Herpotrichiellaceae (Chaetothryiales) obtained from some forest soil samples in Canada using bait plants. *Mycoscience*, 48: 274-281.

Navarro, L. M. (2008). Selección de hospedantes de la cepa DSE1 (Dark Septate Endophyte) y efecto sobre el crecimiento de las plantas infectadas. Thesis, Universidad de Buenos Aires.

Newsham, K. K. (1999). *Phialophora graminicola*, a dark septate fungus, is a beneficial associate of the grass *Vulpia ciliata* ssp. *ambiqua*. *New Phytologist*, 144: 517– 524.

Ocampo, J. A. and Barea, J. M. (1985). Effect of carbamate herbicides on VA mycorrhizal infection and plant growth. *Plant Soil*, 85: 375-383.

Ocampo, J. A. (1986). The effect of VA mycorrhizal infection of host and non-host plants on the growth responses of the plants and competition between them. *Soil Biology and Biochemistry*, 18: 607-610.

Oehl, F., Silva, G. A., Goto, B. T., Sieverding, E. (2011). *Glomeromycota*: three new genera and glomoid species reorganized. *Mycotaxon*, 116:75–120.

Peterson, R. L., Massicotte, H. B., Melville, L. H. *Mycorrhizas; Anatomy and Cell Biology*. Ottawa: NRC Reseach Press. 2004.

Peterson, R. L., Wagg, C., Pautler, M. (2008). Associations between microfungal endophytes and roots: do structural features indicate function? *Botany*, 86: 445-456.

Petrini, O. Taxonomy of endophytic fungi of aerial plant tissues. In: Fokkema, N. J., Van den Huevel, J. (Eds). *Microbiology of the phyllosphere.* Cambridge, UK: Cambridge University Press, 1986, pp. 175–187.

Petrini, O., Sieber, T. N., Toti, L., Viret, O. (1992). Ecology metabolite production and substrate utilization in endophytic fungi. *Natural Toxins*, 1: 185-196.

Pinruan, U., Rungjindamai, N., Choeyklin, R., Lumyong, S., Hyde, K. D., Jones, E. B. G. (2010). Occurrence and diversity of basidiomycetous endophytes from the oil palm, *Elaeis guineensis* in Thailand. *Fungal Diversity,* 41: 1-71.

Ratnayake, M., Leonard, T., Menge, J. A. (1978). Root exudation in relation to supply of phosphorus and its possible relevance to mycorrhizal formation. *New Phytologist*, 81: 543–552.

Reininger, V. and Sieber, T. N. (2012). Mycorrhiza reduces adverse effects of Dark Septate Endophytes (DSE) on growth of conifers. *Plos one*, 7: e42865.

Rodriguez, R. J., Henson, J., Van Volkenburgh, E., Hoy, M., Wright, L., Beckwith, F., Kim, Y. O., Redman, R. S. (2008). Stress tolerance in plants via habitat-adapted symbiosis. *International Society for Microbial Ecology Journal*, 2: 404–416.

Rodriguez, R. J., White, Jr, J. F., Arnold, A. E., Redman, R. S. (2009). Fungal endophytes: diversity and functional roles. *New Phytologist*, 182: 314–330.

Saikkonen, K., Ion, D., Gyllenberg, M. (2002). The persistence of vertically transmitted fungi in grass metapopulations. *Proceedings of the Royal Society B: Biological Sciences,* 269: 1397–1403.

Scervino, J. M., Sampedro, I., Ponce, M. A., Rodriguez, M. A., Ocampo, J. A., Godeas, A. (2008). Rhodotorulic acid enhances root colonization of tomato plants by arbuscular mycorrhizal (AM) fungi due to its stimulatory effect on the pre-symbiotic stages of the AM fungi. *Soil Biology and Biochemistry,* 40: 2474-2476.

Scervino, J. M., Gottlieb, A., Silvani, V. A., Pérgola, M., Fernández, L., Godeas, A. M. (2009). Exudates of dark septate endophyte (DSE) modulate the development of the arbuscular mycorrhizal fungus (AMF) *Gigaspora rosea*. *Soil Biology and Biochemistry*, 41: 1753–1756.

Schulz, B. and Boyle, C. (2005). The endophytic continnum. *Mycological Research,* 109: 661-86.

Schüßler, A. and Walker, C. (2010). The *Glomeromycota*: a species list with new families and genera. Arthur Schüßler and Christopher Walker, Gloucester. Published in December 2010 in libraries at The Royal Botanical Garden Edinburgh, The Royal Botanic Garden Kew, Botanische Staatssammlung Munich, and Oregon State University. Elctronic version freely available online at www.amf-phylogeny.com.

Siddiky, M. R. K., Schaller, J., Caruso, T., Rillig, M. C. (2012). Arbuscular mycorrhizal fungi and collembola non-additively increase soil aggregation. *Soil Biology and Biochemistry*, 47: 93-99.

Sieber, T. N. Fungal root endophytes. In: Waisel, Y., Eshel, A., Kafkafi, U. (Eds.). *Plant Roots: The Hidden Half.* Third Edition ed., New York, Basel: Marcel Dekker, 2002, pp. 887–917.

Smith, S. E. and Read, D. J. (2008). *Mycorrhizal Symbiosis*, third ed. Academic Press, San Diego, 787 pp.

Smith, S. E. and Smith, F. E. (2012). Fresh perspectives on the roles of arbuscular mycorrhizal fungi in plant nutrition and growth. *Mycologia*, 104: 1–13.

Soca-Chafre, G., Rivera-Orduña, F. N., Hidalgo-Lara, M. E., Hernandez-Rodriguez, C., Marsch, R., Flores-Cotera, L. B. (2011). Molecular phylogeny and paclitaxel screening of fungal endophytes from *Taxus globosa*. *Fungal Biology*, 115: 143-156.

Stone, J. K., Polishook, J. D., White, J. F. Jr. Endophytic fungi. In: Mueller, G., Bills, G., Foster, M. (Eds.). Measuring and Monitoring Biodiversity of Fungi. *Inventory and monitoring methods*. Boston, MA: Elsevier Academic Press, 2004, pp. 241-270.

Summerbell, R. C. (2005). Root endophyte and mycorrhizosphere fungi of black spruce, *Picea mariana*, in a boreal forest habitat: influence of site factors on fungal distributions. *Studies in Mycology*, 53: 121–145.

U'Ren, J. M., Lutzoni, F., Miadlikowska, J., Arnold, E. (2010). Community analysis reveals close affinities between endophytic and endolichenic fungi in mosses and lichens. *Microbial Ecology*, 60: 340-353.

Usuki, F. and Narisawa, K. (2007). A mutualistic symbiosis between a dark, septate endophytic fungus, *Heteroconium chaetospir*a, and a non-mycorrhizal plant, Chinese cabbage. *Mycologia*, 99:175–184.

Vaz, A. B. M., Mota, R. C., Bomfim, M. R. Q., Vieira, M. L. A., Zani, C. L., Rosa, C. A., Rosa, L. H. (2009). Antimicrobial activity of endophytic fungi associated with Orchidaceae in Brazil. *Canadian Journal of Microbriology,* 55: 1381–1391.

Vaz, A. B. M., Sampedro, I., Siles, J. A., Vasquez, J. A., García-Romera, I., Vierheilig, H., Rosa, C. A., Ocampo, J. A. (2012). Arbuscular mycorrhizal colonization of S*orghum vulgare* in presence of root endophytic fungi of *Myrtus communis*. *Applied soil ecology*, 61: 288-294.

Whiteside, M. D., Garcia, M. O., Treseder, K. K. (2012). Amino Acid Uptake in Arbuscular Mycorrhizal Plants. *PLoS ONE*, 7: e47643.

Wilcox, H. E. and Wang, C. J. K. (1987). Mycorrhizal and pathological associations of dematiaceous fungi in roots of 7-month-old tree seedlings. *Canadian Journal of Forest Research*, 17: 884-899.

In: Sorghum
Editor: Patricia C. Parra

ISBN: 978-1-62618-351-3

Chapter 5

BREEDING FOR ABIOTIC STRESSES IN SORGHUM

Tariq Shehzad and Kazutoshi Okuno *
Graduate School of Life and Environmental Sciences,
University of Tsukuba, Tsukuba, Japan

ABSTRACT

Mapping and analysis of QTL controlling tolerance to abiotic stresses are primary and significant approaches for germplasm enhancement in crops. In particular, corresponding to global climate changes, improved tolerance to abiotic stresses contributes to crop adaptation and production under stress conditions in crops. The tolerance to abiotic stresses is frequently controlled by multiple genes or quantitative trait loci (QTLs). Identification of QTL using DNA markers is the first strategy in DNA marker-assisted selection (MAS) for the tolerance in sorghum breeding. We assessed salt and drought tolerances using a core collection derived from world germplasm collections of sorghum and detected a broad range of variation. Association analysis was performed using data on genotypes at 98 SSR marker loci and phenotypic values. Major QTL for salt tolerance was mapped on chromosome 1 and closely linked to QTL for leaf senescence. Nine QTL for drought tolerance were also detected. Based on the genetics of

* Corresponding author: Prof Kazutoshi Okuno. Email; okuno.kazutoshi.fu@u.tsukuba.ac.jp. Phone: +81-29853-4807; Fax: +81-29853-6617.

tolerance to abiotic stresses, we discussed breeding strategies for abiotic stresses in sorghum.

1. INTRODUCTION

Sorghum [*Sorghum bicolor* (L.) Moench] is an important food and feed crop in many parts of the world. Recently sorghum is getting more attention to be used as biofuel crop. Its remarkable ability to grow under adverse conditions, in particular with much less water than most other grain crops, makes sorghum an important "failsafe" source of food, feed, fiber, and fuel in the world. Sorghum is more tolerant to many abiotic stresses, including heat, drought, and flooding, making it an ideal crop for growing on marginal lands as demands for food, feed, and energy increase. As a model crop for "C4" photosynthesis tropical grasses it substitutes a C3 rice in comparative genomic studies of more complex grasses such as maize and sugarcane. Sorghum has a very diverse germplasm, and molecular markers are being used to assess this diversity to help manage large germplasm collections make these collections more useful in genetic dissection of important agronomic traits.

Climate change is thought to be a major source of abiotic stress for crops. Changing patterns of temperature and rainfall are having an impact on the cultivation of food and other crops, with some formerly productive areas suffering from drought, flooding or extremes of temperature. In order to alleviate the economic impact of crop failures, research is being carried out into the development, through breeding or genetic engineering, of crop plants that are more resistant to these forms of stress.

With a still rising world population, pressure on agriculture to produce more food is increasing. High-producing crops are therefore very much in demand. Abiotic stresses like for instance drought, salinity, and temperature stress account for most of the crop losses each year. Crop yields considerably in the last 50 years by optimizing environmental conditions for agriculture and selection for high yielding plants, but it puts an enormous strain on our natural resources. For instance, 70% of the world's fresh water supplies are used for agriculture. In addition, agricultural land is claimed more and more for bio-energy production, which increases the need to be able to grow crops on more marginal grounds.

Breeding for Abiotic Stress Tolerance strives to contribute to our knowledge on the genetics underlying tolerance to abiotic stresses, in particular to drought and salinity. Increased understanding of the genetics of the plant's response to drought and salinity will enable focused breeding for increased tolerance to these abiotic stresses which will ultimately lead to crops that consistently give high yields even on marginal land and when experiencing drought or salinity stress.

The use of mapping and segregating populations, cultivar and diverse germplasm collections to assess the natural variation in different crops including wheat, barley, rice and sorghum for drought and salinity tolerance has enhanced breeding improvement in sorghum. The focus is to investigate transpiration efficiency, relative water content, chlorophyll fluorescence, chlorophyll content, Na+ and K+ content, root growth and architecture etc. to assess stress tolerance in crops.

2. Using Molecular Makers

Molecular markers an essential tool for genomic and genetic research. Molecular markers allow discrimination in DNA sequence amongst cultivars and breeding lines and thus offer the scientific community singularly powerful tools to monitor, track, and exploit sequence variation in germplasm. Many types of markers have been developed and are now an essential part of functional, structural genomics and molecular breeding. Different kinds of molecular makers are being used for the dissection of important traits directly or indirectly related to stresses in crops. Microsatellite markers or single sequence repeats (SSR) have proved useful in wheat research since they offer reproducibility, multiallelic nature, codominant inheritance, genome specificity, relative abundance, and good genome coverage. Another promising type is single-nucleotide polymorphism (SNP) or biallelic markers because they form the basis for most genetic variation between individuals; they are widely distributed and amenable to high multiplex detection systems. The identification of SNP markers depends on comparative sequencing of lines or analysis of expression sequence tags (ESTs). This was delayed in wheat due to a lack of adequate technologies that can handle the problems associated with polyploidy and sequencing large genomes. Molecular markers can be derived either from within or from outside the gene sequence and the selection of the most suitable marker system depends on the objective and cost.

3. Quantitative Trait Loci (QTL) Discovery

The advances in functional genomics in delivering bacterial artificial chromosomes (BACs), ESTs, partial gene sequences, full-length cDNA clones, genes and markers have enabled the establishment of molecular maps based on genic markers. The creation of suitable mapping populations and the development of molecular markers have enabled linkage studies in wheat and many QTLs have been identified for yield under drought environments (Varshney *et al.*, 2006). Linkage studies have shown that QTLs for grain yield reside in several chromosomal regions, and measurements of yield components allow us to dissect complex traits to smaller genetic components more amenable for building our knowledge of trait architecture, which will inform our future strategies for exploitation. Mapping populations have also been developed in durum wheat for the study of biotic and abiotic stresses (Blanco *et al.*, 1998; Nachit *et al.*, 2001; Nachit and Elouafi, 2004). Despite the available maps, populations, and marker technology, advances in transferring knowledge from QTL studies on yield under drought to breeding remains slow.

This is due to three major factors: first, yield is a quantitative trait with poorly defined genetic architecture of multiple interacting loci, the second is the fact that these loci have a high *G*×*E* interaction component, and the third is the accurate phenotyping of the traits under study. QTLs established in one environment often disappear in another and thus QTLs have been defined as constitutive versus adaptive. However, this categorization can be artificial since certain genes underlying ‘constitutive’ traits can also respond to the environmental conditions such as flowering genes (Collins *et al.*, 2008).

To enable a study of the interaction of *G*×*E*, phenotyping trials can be carried out on the same population in multiple field sites where environmental covariates are considered in the analysis. To overcome this problem, I would like to grow the same population in multiple environments and taking advantage of research collaborations with multiple groups. This allows the simulation of various gene actions or models in the context of selection and breeding strategies.

Despite the complexity of studying yield and its genetic architecture under drought and salt stress, data are now available from multiple studies and from various genetic materials to enable us to start identifying important loci for further work.

3.1. Exploring and Fine Mapping a QTL

Primary QTL mapping will identify a QTL to within 10–30 cM resolution. To fine map a locus to a higher resolution, near isogenic lines (NILS) specific for the particular QTL under study are required to Mendelize the locus (Alonso-Blanco and Koornneef, 2000). Only a handful of studies have achieved the cloning of a wheat QTL via positional cloning, most notably the cloning of *VRN1* (Yan *et al.*, 2003). When a QTL region is resolved to a few cM genetic distance, markers closest to the QTL are then used to anchor the genetic map to a physical map and BAC libraries are then used to identify potential candidate sequences. The use of model species, synteny, and bioinformatics tools is critical at this stage to enable candidate genes to be identified in wheat. Whilst QTL cloning represents a huge undertaking in terms of the technology, time, and resources required, the advantages derived from its success are directly applicable to gene cloning.

4. Candidate Gene and Genetic Engineering Approach

High throughput transcriptome studies and functional genomic approaches have been used to identify genes correlated with a plant's response to water stress. A large number of genes have been identified and the diversity of these responsive genes and pathways reflects the complexity of the mechanisms involved in sensing and responding to water stress. These genes have been categorized by Yamaguchi-Shinozaki and Shinozaki (2006) into two major groups: the first codes for functional proteins involved in protecting cellular function (water channels, transporters, detoxification, proteases, protection factors of macromolecules, chaperones, and osmolyte biosynthesis); the other group codes for regulatory proteins involved in signal transduction such as transcription factors, protein kinases, protein phosphatases, and enzymes of lipid metabolism (Shinozaki *et al.*, 2003). Genetic modification of the expression of members belonging to these two major groups of genes, further categorized as abscisic acid (ABA)-dependent or ABA-independent, has resulted in some improvement in traits relating to tolerance under water stress in a variety of plant species. However, only a few genes have been successfully engineered in crops to enhance resistance to drought under field conditions. An important aspect that is worth highlighting here is that fine-tuning the modulation of the target gene(s) expression to specific cell types may be essential in a successful strategy for modulating the system.

5. Use of Core Collection in Genetic Analysis

A core collection is a limited set of accessions representing, with a minimum of repetitiveness, the genetic diversity of a crop species and its wild relatives (Frankel 1984). Core collections have been commonly used to screen for agronomically useful traits or characteristics. Such collections proved as powerful tools to study the genetic mechanisms of important physio-chemical processes in crops. Similarly the extent of either genotype by environment (G × E) interactions or gene or QTL by environment interactions can be asssessed and exploited in a genepool (Charmet *et al.* 1993).

We are working to exploit natural variation in germplasm to enhance salt tolerance in important crops including sorghum. We have developed microsatellite or simple sequence repeat (SSR) markers based mini core collection of 107 sorghum accessions selected from Asian and African regions of the globe (Shehzad *et al.,* 2009a). The representativeness of the collection can be judged by comparing with original population. The core set developed in this study retained 99.93 % of total alleles present in whole population. Similarly, no drastic change was observed for genetic diversity attributes like % polymorphic loci, similarity coefficient and gene diversity regarding base collection. Based on our results, we were able to construct a diversity research set of 107 accessions from 320 diverse geographic origins of Africa and Asia i.e. 33 % of the total number. A wide range of variation can be observed for different important agronomical and physiological traits in the germplasm including panicle shape ranging from loose to very compact type, as shown in Figure 1. In the assessments of the 98 SSR markers, a total of 470 alleles were observed and the alleles could uniquely classify 107 sorghum accessions into three sub populations by structure analysis. The first population contained 33 sorghum accessions mostly from Africa origin while the second population contained 36 accessions, from both Africa and Asia regions. The third population contained 38 accessions mostly from East Asian origin (Shehzad *et al.* 2009b). In general, conventional method for QTL analysis in germplasm is linkage mapping. Identifying QTLs by linkage mapping needs to construct one or several segregating populations by crossing between parents (e.g., F_2, Double haploid, Backcross population). Linkage analysis depends upon recent genetic recombination between two different plant lines (as the result of a genetic cross) to identify general regions of interest, with the advantage of requiring few genetic markers to ensure genome-wide coverage and high statistical power per allele. However, it has the disadvantages of low mapping resolution and low allele richness. An alternative, association mapping based on linkage disequilibrium (LD) analysis might be an effective way to identify the function of the gene (or targeted high-resolution QTL). The basic scheme of association mapping is given in Figure 2. We are utilizing both approaches i.e. linkage mapping as well as association mapping dissect the mechanism of response of sorghum to abiotic stresses and identification of chromosomal regions controlling such characteristics.

Figure 1. Various panicle types / head pattern ranging from very open and loose to very compact in sorghum core collection.

6. Salt Tolerance in Sorghum

Soil salinity is one of the major abiotic stresses drastically affecting agricultural productivity (Boyer, 1982). The increase in levels of saline areas affects large terrestrial areas of the world. It is estimated that more than one third of irrigated land in the world is presently affected by salinity. This is exclusive of the regions classified as arid and desert lands (which comprise 25% of the total land of our planet). The loss of farmable land due to salinization is directly in conflict with the needs of global population expected to reach between 7.5 and 10.5 billion by 2050.

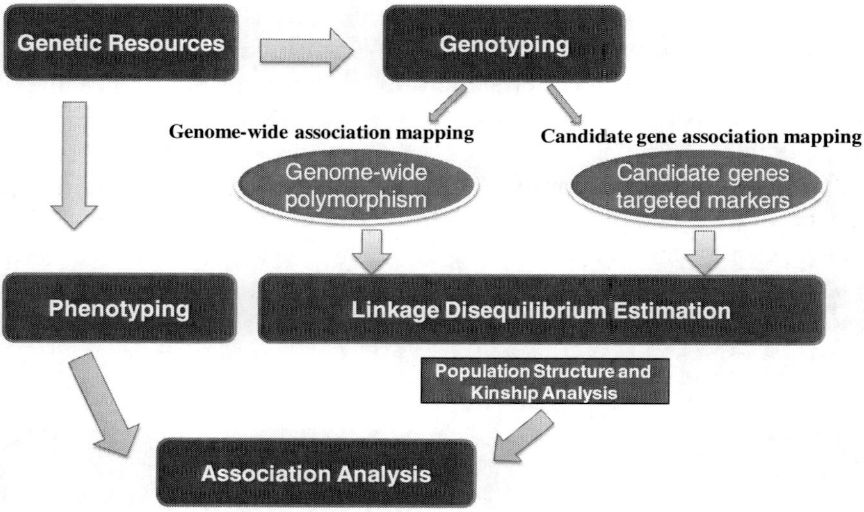

Figure 2. General Scheme of association mapping.

Two basic genetical approaches currently being utilized to improve salt stress tolerance in plants including (1) exploitation of natural genetic variations, either through direct selection in stressful environments or through the mapping of QTLs and subsequent marker-assisted selection and (2) generation of transgenic plants to introduce novel genes or alter expression levels of the existing genes to affect the degree of salt stress tolerance. Plant species differ in their salt tolerance depending on their genetic makeup ranging from high to low levels of salts in the soil. Sorghum is considered as moderately tolerant to salinity and can be grown in saline soils. Salinity is the most important environmental factor affecting the crop production in many parts of the world. Almost 19.5% of all irrigated land is affected with salinity and 2.1% of dry-land agriculture worldwide (FAO 2005). Salinity reduces the ability of plants to utilize water and causes a reduction in growth rate, as well as changes in plant metabolic processes (Munns 2002).

6.1. Phenotypic Assessment of Salt Tolerance

Scientists have made consensus about the importance of phenotyping methodologies and consider it as the most limiting factor in breeding for abiotic stress tolerance in crops. Advanced phenotyping to characterize plant architecture and performance, has the potential to accelerate breeding progress but is far from being routinely used in current breeding approaches. In breeding populations with high genetic diversity and substantial G x E interactions, precise and efficient phenotyping is essential to meet future challenges imposed by climate change, growing demand and declining resources. We have screened our sorghum core collection for salt tolerance in pot as well as hydroponic experiments (unpublished data). Preliminary we used 0, 150, 200 and 250 mM concentrations of salt and 250mM was found to be the most limiting with low coefficient of variation (CV%) within genotypes. While comparing different methods of phenotyping, we have fixed developed a new system for the assessment of salt tolerance as shown in Figure 3. In this system the environmental variance is controlled at maximum level providing constant and homogenous exposure of plants to saline water. Different morphological traits directly or indirectly related to salt tolerance assessments were studied but mortality rate (Mortality rate = number of withered leaves / total number of leaves x100) was found to be the most determinant trait with less than 10% CV within accession.

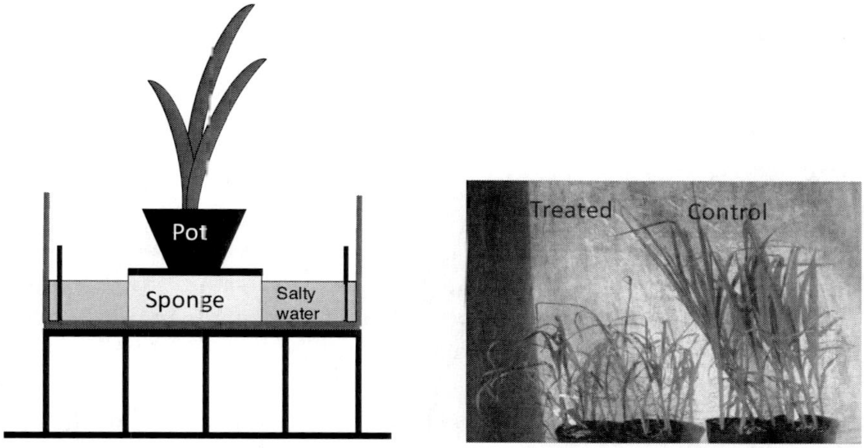

a b

Figure 3. Evaluation of salt tolerance in sorghum a) method of phenotypic assessment of salt tolerance b) response of a sorghum accession to salt treatment (unpublished data).

6.2. Mapping QTLs of Salt Tolerance

Due to the complex nature of salt tolerance, the large number of genes implicated in salinity tolerance may not be surprising, because many of the traits underlying adaptation to abiotic stress are quantitative and controlled by more than a single gene with lots of genetic pathways (Sreenivasulu *et al.* 2007). We assess natural variation for salinity tolerance in *Sorghum bicolor* and utilize both association as well as QTL mapping approaches to identify QTLs that would leads towards cloning of candidate genes. Because of the physiological complexity of the salinity response, there have been few studies so far reported for the identification of chromosomal regions controlling this important trait.

In our preliminary studies we used association mapping approach to identify QTLs responsible for salt tolerance in sorghum. For the traits related to salt tolerance, four QTLs were detected including two QTLs for plant height (each on chromosomes 2 and 6) were detected whereas, two QTLs for dry weight on chromosomes 4 and 6, respectively. Similarly, one major QTL was identified on chromosome 1 closely linked the position of the gene controlling leaf senescence in crops (unpublished data). Furthermore, we are developing different segregating populations to dissect the loci responsible for salt tolerance in sorghum by linkage analysis.

7. Drought Tolerance in Sorghum

Adaptation to drought conditions is one of the most complex biological processes. It involves numerous pathways including transcriptional activation/inactivation of specific genes, transient increases in ABA levels, accumulation of compatible solutes and protective enzymes, increased levels of antioxides, suppression of energy-metabolism pathways etc. Drought significantly reduces plant productivity by inhibiting growth and photosynthesis (Taiz *et al.* 1998). In plants atleast four independent stress responsive genetic regulatory pathways are known to exist, forming a highly complex and redundant gene network (Shinozaki *et al.* 2007; Umezawa *et al.* 2006). Two of the pathways are ABA dependent whereas, two are ABA independent. These pathways are also linked to other biological processes involved in other stress responses, including cold, high temperature, and salinity. Many studies on drought tolerance monitored the physiological and biochemical status of stressed plants compared with unstressed plants.

Detection of quantitative trait loci (QTLs) for drought tolerance is the first step of genetic improvement to stabilize global crop production. Pre-flowering drought stress is the most crucial stage affecting many important physiological processes at different developmental stages. The effects of pre-flowering drought on developmental growth stages are given in the schematic diagram of Figure 4.

7.1. Phenotypic Assessment of Drought Tolerance

We have screened our core collection for 23 different drought related traits including photoperiod sensitivity, yield components and other related traits. However leaf drying score (LDS) was used as a parameter for the assessment of drought tolerance/sensitivity ranging from "0" as extremely tolerant to "9" as extremely sensitive genotypes. Leaf drying was visually scored by calculating leaf drying percentage as;

Dry leaves percentage = (number of all dry leaves/total number of leaves) x 100.

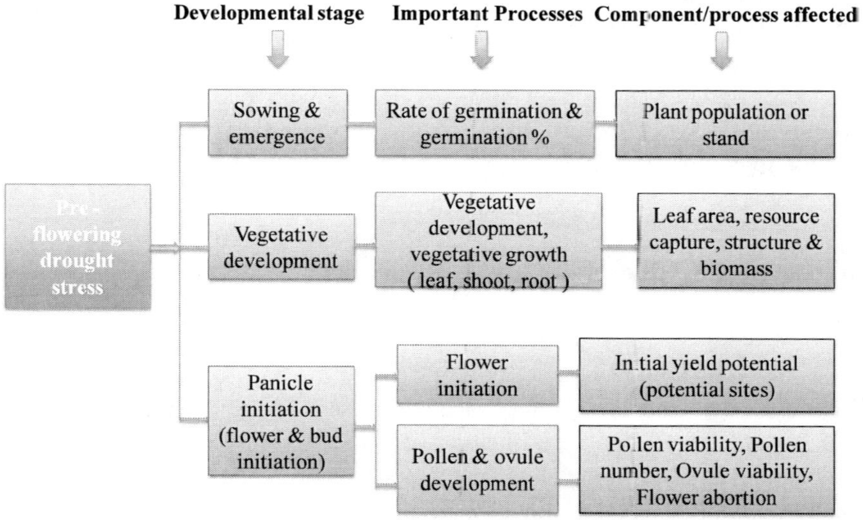

Figure 4. Effects of Pre-flowering drought stress on different important growth stages.

Figure 5. Response of some sorghum accessions to drought stress vs. control (unpublished data).

According to the development of scale on basis of dry leaves percentage, each plant corresponded to the scale in the following way: Scale-0= no dry leaves (not sensitive), 1= 1-10 % (extremely low), 2= 11-20 % (very low), 3= 21-30 % (low), 4= 31-40 % (slightly low), 5= 41-50 % (intermediate), 6= 51-60% (slightly high), 7= 61-70% (high), 8= 71-80% (very high), 9= 81-90% (extremely high) or more than 90% dry leaves with plant death symptoms (unpublished data). The response of sorghum accessions to drought stress condition in comparison to control is presented in Figure 5.

7.2. Mapping QTLs of Drought Tolerance

Drought tolerance in plants is quantitative in nature, and controlling by polygenes. Association mapping is an effective method to identify QTLs for this complex mechanism. We have also applied association mapping approach in our germplasm for the identification of QTLs. To minimize false positive and negative rates, different statistical models of association analysis were compared. Finally Q-model (model with population structure) and Q+K model (model with population structure and kinship) were selected. Under controlled condition 17 QTLs were identified on chromosomes 1, 2, 4 8, 9 and 10 with *p*-values ranged from 2.5 to 7.6 explaining 9.5 to 57.5% of total phenotypic variance. Under drought-stressed condition 9 QTLs were identified explaining 9 to 61.2% of total phenotypic variance with *p*-values ranged from 2.5 to 3.5 on chromosomes 1, 2, 3 and10 (unpublished data). The QTLs identified in this study on chromosomes 1, 2, 3 and 10 showed consistency with the previously identified QTLs for stay green traits under drought stress. Similarly, the major QTL identified for LDS on chromosome 1, is same as previously mentioned for salt tolerance in sorghum. Two QTLs identified were physically localized on chromosome 2 and were homologous for genes encoding DNA binding and ATP binding proteins. These QTLs could be used to dissect further the genetic and physiological basis of drought tolerance in sorghum.

FUTURE PROSPECTS

Conventional breeding programs for abiotic stress tolerance have achieved limited success. This lack of success is in part because breeders prefer to evaluate their genetic material in ideal conditions. There is a need to develop more high throughput phenotyping system for salt and drought assessments. The statistical power of linkage and association analyses needs to be enhanced by reducing type I and type II errors. Use of advanced molecular techniques will further promote the identification of gene and gene networks for abiotic stresses in crops. While it has been recognized by many researchers that there are dramatic changes in gene expression associated with all types of stresses. We also need to establish better comparative systems for genomic studies.

There are a number of research areas, which are still at improvement phase and will help to find answers of many fundamental questions associated with stress tolerance. Such as development of molecular markers for gene mapping and development high throughput linkage maps, development of EST libraries, comparative genomic studies including Arabidopsis, rice and sorghum, and development of TILLING (Targeting Induced Local Lesions IN Genomes) techniques.

REFERENCES

Alonso-Blanco, C., Peeters, A.J., Koornneef, M., Lister, C., Dean, C., van den Bosch, N., Pot, J., & Kuiper, M.T. (1998). Development of an AFLP based linkage map of L*er*, Col and Cvi *Arabidopsis thaliana* ecotypes and construction of a L*er*/Cvi recombinant inbred line population, *Plant J* 14, 259–271.

Blanco A., Bellomo, M.P., Cenci, A., et al. (1998). A genetic linkage map of durum wheat, *Theor and Appl Genet* 97, 721–728.

Boyer, J.S. (1982). Plant productivity and environment, *Science* 218, 443–448.

Charmet, G., Balfourier, F., Ravel, C., & Denis, J.B. (1993). Genotype x environment interactions in a collection of French perennial ryegrass populations, *Theor Appl Genet* 86, 731–736.

Collins N.C., Tardieu, F., & Tuberosa, R. (2008). Quantitative trait loci and crop performance under abiotic stress: where do we stand? *Plant Physiology* 147, 469–486.

FAO, (2005). Global network on integrated soil management for sustainable use of salt affected soils, Rome, Italy: FAO Land and Plant Nutrition Management Service.

Frankel, O.H. (1984). *Genetic perspectives of germplasm conservation.* Pp. 161-170 in Genetic Manipulation: Impact on Man and Society (W. Arber, K. Llimensee, W.J. Peacock and P. Starlinger, eds.). Cambridge University Press, Cambridge.

Munns R. (2002). *Comparative physiology of salt and water stress,* Plant Cell Environ 25, 239–250.

Nachit M.M., & Elouafi, I. (2004). Durum adaptation in the Mediterranean dryland: breeding, stress physiology, and molecular markers. In: Rao SC, Ryan J, eds. *Challenges and strategies for dryland agriculture.* CSSA Special Publication 32. Madison, Wisconsin, USA: Crop Science Society of America Inc., American Society of Agronomy Inc, 203–218.

Nachit M.M., Elouafi, I., Pagnotta, M.A., et al. (2001). Molecular linkage map for an intraspecific recombinant inbred population of durum wheat (*Triticum turgidum* L. var. durum), *Theor and Appl Genet* 102, 177–186.

Shehzad, T., Iwata H., & Okuno, K. (2009b). Genome-wide association mapping of quantitative traits in sorghum (*Sorghum bicolor* (L.) Moench) by using multiple models, *Breeding Science* 59, 217–227.

Shehzad, T., Okuizumi, H., Kawase, M., & Okuno, K. (2009a) Development of SSR-based sorghum (*Sorghum bicolor* (L.) Moench) diversity research set of germplasm and its evaluation by morphological traits. *Genet Resour Crop Evol* 56, 809-827.

Shinozaki K., & Yamaguchi-Shinozaki K. (2007). Gene networks involved in drought stress response and tolerance, *J Exp Bot* 58, 221–227.

Shinozaki K., Yamaguchi-Shinozaki, K., & Seki, M. (2003). Regulatory network of gene expression in the drought and cold stress responses, *Curr Opin Plant Biol* 6, 410–417.

Sreenivasulu N., Sopory S.K., & Kavi Kishor P.B. (2007). Deciphering the regulatory mechanisms of abiotic stress tolerance in plants by genomic approaches, *Gene* 388, 1–13.

Taiz, L., & Zieger, E. (1998). Stress Physiology. In: *Plant Physiology, 2nd edn. Sinauer Associates,* Inc., Sunderland, M.A. pp. 725-757.

Umezawa T., Fujita M., Fujita Y., Yamaguchi-Shinozaki K., & Shinozaki K (2006). Engineering drought tolerance in plants: discovering and tailoring genes unlock the future, *Curr Opin Biotech* 17,113–122.

Varshney, R.V., Hoisington, D.A., & Tyagi, A.K. (2006). Advances in cereal genomics and applications in crop breeding, *Trends Biotechnol* 24, 490–499.

Yamaguchi-Shinozaki, K., Sakuma, Y., Ito, Y., & Shinozaki, K. (2006). The DRE/DREB regulon of gene expression in Arabidopsis and rice in response to drought and cold stress. In Ribaut, J.M. (ed) *Drought Adaptation in Cereals.* The Haworth Press, New York, 583–598.

INDEX

A

B

C

F

G

H

I

N

O

P

Q

R

S

T

U

V

W

Y

Z